Remote Sensor Monitoring by Radio with Arduino: Detecting Intruders, Fires, Flammable and Toxic Gases, and Other Hazards at a Distance

By

David Leithauser

Copyright © 2017
David Leithauser

All right reserved

Table of contents

Introduction 3

Hardware Aspects of Using the nRF24L01 4

Installing Basic Software for the nRF24L01 12

Software for the nRF24L01 Transmitter 14

Software for the nRF24L01 Receiver 23

Using Repeaters to Extend Your Transmission Range 51

Detecting Intruders 57

Fire 79

Temperature Extremes 84

Toxic and Flammable Gases 92

Flooding or Too Little Water 106

Power Failure or Low Power 108

Other Sensors 117

Downloading files and Contacting the Author 118

Introduction

This book is about connecting sensors and radio transceivers to an Arduino so that you can monitor the sensor readings from a distance. You can put the Arduino-sensor package miles away from the receiving station, in your front or back yard, or even in your home like your basement or attic.

Although the techniques described in this book will work with any type of sensor input, the book will focus on sensors that detect potentially dangerous or disruptive conditions. These will include intruders, fires, flammable gas leaks and other toxic gases like pollution, power failures, floods (including minor "floods" like a pipe bursting), and other hazards.

For the radio communications, we will use the nRF24L01 transceiver chip. This inexpensive chip (usually around $1.00 on EBay) interfaces easily with the Arduino and can both transmit and receive data. It has an advertised range of 100 meters (about 328 feet) for the basic unit, although in actual practice it may be closer to 30 meters (about 98 feet). However, with an optional antenna the range is reported to be 1,000 meters (1 km, about .6 miles). In the first five chapters, I explain the hardware and software aspects of this handy transceiver, enabling you to set up the communications. I even explain how to set up repeater transmitters that can relay the signal from locations beyond the 1 km range. In the chapters after these five chapters, I discuss attaching and operating various sensors, explaining how to set them up and integrate them into the transmission software. The chapters will be divided by hazards you can monitor, not specific sensors, so one chapter may include several different types of sensors that can be used to detect the same hazard.

Chapter 1

Hardware Aspects of Using the nRF24L01

The basic and simplest version of the nRF24L01 transceiver is shown from the top in Figure 1.1 and the bottom in Figure 1.2, and the unit with the range extending antenna is shown in Figure 1.3.

Figure 1.1

Figure 1.2

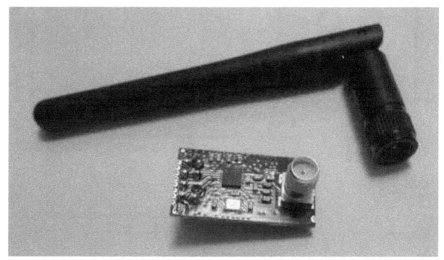

Figure 1.3

There are eight pins to connect the unit to the Arduino, although one is an interrupt that is not really necessary and is seldom used, so we will not discuss it in this book. A sketch of the pin configuration, as seen from the back with the pins pointed toward you, is shown in Figure 1.4. This is the way you will see it when you attach the connectors to the pins. Note that many diagrams you see in the Internet show the pins as seen from the top of the unit, which is confusing because that is not the way you would be looking at it when you are connecting the cables to the unit.

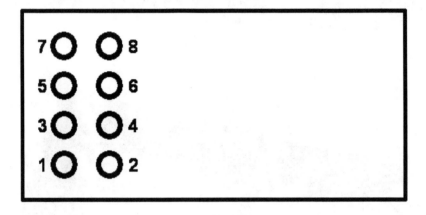

Figure 1.4

Using this diagram, the function of each pin and the Arduino pin you should connect it to (on an Arduino Uno) is shown in Table 1.1

nRF24L01 pin	Function	Arduino pin
1	Ground	GND
2	3.3 V power	3V3
3	CE	D9
4	CSN	D10
5	SCK	D13
6	MOSI	D11
7	MISO	D12
8	IRQ (interrupt)	none

Table 1.1

The CE and CSN Arduino connection is actually configurable within the sketch itself. That is, there is a line of code in the sketch that tells the Arduino which pins the CE and CSN pins of the nRF24L01 are connected to, so you can change these if for some reason you absolutely need the Arduino D9 and D10 pins for something else. Some sketches you will find on the Web use other pins, such as D7 and D8, so be careful if you try to combine sketches from different sources. I will use D9 and D10 throughout this book for consistency.

It is VERY important to note that the positive power supply for the nRF24L01 is the Arduino 3.3 volt outlet, NOT the 5 volt outlet. Connecting the nRF24L01 to 5 volts will fry it. There is a way to connect the nRF24L01 to the Arduino 5 volt power pin, if you really want to. You can buy an attachment known as a socket adapter plate board that contains a voltage regulator that will drop the 5 volts to 3.3 volts and also smooth out the current. This device is shown in Figure 1.5.

Figure 1.5

The nRF24L01 plugs into the socket and you can then connect cables to the pins on the socket adapter plate board to your Arduino. The pins are labeled so you can use Table 1.1 to see which pins on the socket adapter plate board to connect the proper pins on the Arduino. On the socket adapter plate board, MOSI is abbreviated MO and MISO is abbreviated MI, but otherwise the markings match Table 1.1. Note that VCC on the socket adapter plate board goes to 5V on the Arduino.

There have been some complains that data is sometimes dropped during transmission due to power fluctuations when the nRF24L01 is powered by the Arduino 3.3 volts. Some people like to solder a small capacitor, usually about 3.3 or 4.7 microfarads, to the power pins of the nRF24L01 to stabilize the power. This is shown in Figure 1.6.

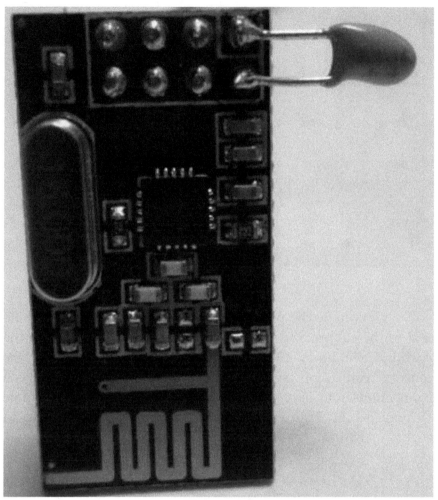

Figure 1.6

To connect the nRF24L01 to the Arduino, you can use male-female connector cables. These are generally referred to a Dupont cables on EBay or Amazon.com.
They are shown in Figure 1.7.

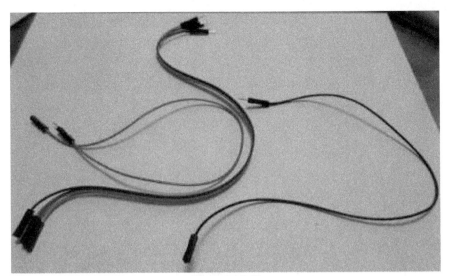

Figure 1.7

Of course, the pins on the nRF24L01 go into the female end of the cables, and the male pins on the cables go into the Arduino pin holes. On some models of Arduino, there are pins as well as pin sockets, allowing both male and female connections. On such Arduinos, you can use female-female Dupont cables. Figure 1.8 shows the nRF24L01 connected to such an Arduino.

Figure 1.8

    I actually recommend using these Arduino BUONO and Arduino UNO LC units. They have the following advantages:

1) They provide more power (current) to the 3.3 volts and 5 volt power outlets.

2) The Uno LC has both pin and pin hole connections, providing twice as many connections

3) Connecting a female-female Dupont cable to the Arduino pins provides a stronger and more secure connection than inserting a pin or wire into the Arduino pin hole connection.

Chapter 2

Installing Basic Software for the nRF24L01

Before you begin writing any code for the nRF24L01 transceiver, you will need to install the necessary preliminary software. Of course, you must first install the Arduino IDE software, which allows you to write programs for the Arduino, compile them into machine language, and load them onto your Arduino. I assume in this book that you are somewhat familiar with programming the Arduino in general, but just in case a few readers are completely new, I will mention that you can download the IDE from

https://www.arduino.cc/en/Main/Software

BTW, the version of the Arduino IDE used for writing this book was 1.6.5.

To use the nRF24L01, you must download the libraries for it. Here are a few Web pages where you can download this:

https://github.com/maniacbug/RF24/

Click on the "Clone or download" button on this page and then select "Download ZIP" to download the archive to your computer. The file that downloads should be called RF24-master.zip.

I have also put this on my own Web site and my public Box folder. The links to directly download the file are
http://LeithauserResearch.com/RF24-master.zip
and
https://app.box.com/s/xsf7e6gsof5jteijvcq7zd38cfwcu53r

After downloading the file, run the Arduino IDE. Click on the "Sketch" menu at the top, then the "include library" option on the menu that drops down, then on "Add ZIP library." In the files and folder box that appears, navigate to where you saved the RF24-master.zip file and double click on the RF24-master.zip file. That should install the library.

You can find many more just by doing a search for RF24 download on Google or other search engines. However, I have deliberately selected these libraries because they are no longer being updated, so there is no chance of the authors making changes in them that will make them incompatible with this book. If you use other libraries, you may suddenly find them generating errors when you try to compile the sketches in this book.

Note: Do not add more than one library for the nRF24L01. Doing so can cause the Arduino IDE to pull library files several different libraries, causing errors.

Chapter 3

Software for the nRF24L01 Transmitter

Now that we have discussed the hardware and basic software setup, we can get onto the software. This chapter will discuss the code (or sketch) for the transmitter. Note: In this book, I will be using the similar terms sketch, program, code, and software interchangeably. These terms mean essentially the same thing, with very slight differences, such as the fact that code can refer to a small section of code while sketch usually means the entire sketch.

First, we need to include the libraries with this code:
#include <SPI.h>
#include <RF24.h>
Some authors add the code
#include <nRF24L01.h>
at the beginning, but I have found that it works fine without this. I suspect that this is some vestigial code that was once necessary and people kept putting it in after it no longer was necessary because they were unaware that it is no longer necessary.

Next we can define the Arduino location of the CE and CSN pins.
#define CE_PIN 9
#define CSN_PIN 10

Then we inform the Arduino where these pins are with this statement
RF24 radio(CE_PIN,CSN_PIN);
Note that we could have simply written
RF24 radio(9,10);
However, this way better documents the code and makes it easier in case we want to change the pins later.

Next we have some variables and constants to define. The most important is a pipe for the transmission of the data. I will not go into a lot of technical details about that here, but here is the code.

const uint64_t pipe = 0xE8E8F0F0E1LL;

Next some constants and variables specifically for the operation of this sketch.

int dataArray[3]; // Variable to hold data
int success; // variable to see if transmission succeeded
int tries; // Number of tries to transmit
const int maximumTries = 10;

The array dataArray is simply an array to hold the data that is to be transmitted. It does not have to be integers. It can be float. (However, the type in the receiver discussed in the next chapter must match the type in the transmitter.) It can be any number of dimensions, depending on how many sensors you have on your device and how much additional information you want to send. (Again, the size in the receiver and transmitter must match.) I will get into suggestions for some of the other information you might want to send along with your actual data from your sensors, but for now we will just pick 3 as the number of values we will send, and therefore dimension the array as 3. The variable success is used to determine whether transmission has been successful. The variable tries is how many times you have tried to transmit the data, and maximumTries is a number you set for how many times you want the Arduino to keep trying to send the data if it is having trouble. The reason for setting any maximum number of tries at all is simply to prevent the code from looping indefinitely if something is seriously wrong, like the nRF24L01 has gotten disconnected or something.

Next we need to set up the operation with the normal setup routine found in all Arduino sketches. Here is the setup for this sketch.

```
void setup() {
 Serial.begin(9600);
 radio.begin();
 radio.openWritingPipe(pipe);
}// end setup
```

The radio.begin(); line tells the Arduino to activate process, and the radio.openWritingPipe(pipe); line opens (activates) the transmission pipe. The Serial.begin(9600); line is not always necessary in a fully functional Arduino sketch, but it is generally included to help with testing and diagnosis of the program.

Next we come to the actual communications code. For the sake of clarity, we will put this in its own subroutine.

```
void transmitData(){
   success = 0;
   tries=0;
   while (tries<maximumTries && success==0){
   success = radio.write(dataArray, sizeof(dataArray));
   tries = tries + 1;
  }
}
```

The actual command is radio.write. This takes two parameters, the actual array of data and the size of the array. In this sketch, the array is dataArray. There is a useful function in the Arduino library called sizeof that tells you the size of an array automatically. Thus, the command is radio.write(dataArray, sizeof(dataArray)). In the libraries we are using, radio.write is a function that returns 1 if transmission is successful and 0 if it fails. This is one reason I prefer this library. We are putting the result of this return in the variable success.

We set success and tries to 0. The while loop will continue trying to transmit the data until success is no longer 0 (which will happen once transmission is successful) or the number of tries exceeds the limit set by maximumTries. Remember that you can make maximumTries as large as you like, but 10 (as set in this program) should be more than adequate. Once transmission succeeds or you exceed the maximum tries, the while loop terminates and the program flow returns from the transmitData subroutine.

The main loop routine of the sketch will fill the dataArray array with data and then call the transmitData subroutine. There are several ways that you can have the loop decide when to transmit the data. Since we are concentrating on detecting hazards, you can have the sketch transmit the data each time it detects a hazard. If you do this, you might also want to have it transmit if the hazard situation ends. For example, if you are detecting intruders, you might want the sketch to transmit when it detects and intruder and also when it no longer does. If you are monitoring a situation constantly, you might want the loop to simply transmit periodically. For demonstration and testing purposes, I will simply provide some code here that allows you to send some numbers from the serial monitor of the Arduino IDE.

```
void loop() {
  if (Serial.available() > 0) {
    dataArray[0] = Serial.parseInt();
    dataArray[1] = Serial.parseInt();
    dataArray[2] = Serial.parseInt();
    transmitData();
    if (success) {
      Serial.print("Success on attempt ");
    }
    else{
      Serial.print("Failure after attempt ");
    }
    Serial.println(tries);
  }
}// End main loop
```

This waits for input from the serial port. You would type three numbers separated by commas, such as 1,2,3 in the serial monitor. The function Serial.parseInt() converts each number into an integer. Each integer is stored in one of the array elements. The sketch then calls the transmitData() subroutine to transmit this data. Just for testing purposes, the code after transmitData() checks to see if success is no longer 0. If it is, the code serial prints the message "Success on attempt ". If success is still 0, it prints "Failure after attempt ". It then prints the number of tries it made. A typical return message might be "Success on attempt 2."

The entire sketch is shown in Sketch 3.1

```
#include <SPI.h>
#include <RF24.h>

#define CE_PIN 9
#define CSN_PIN 10
RF24 radio(CE_PIN,CSN_PIN);


const uint64_t pipe = 0xE8E8F0F0E1LL;

int dataArray[3]; // Variable to hold data
int success; // variable to see if transmission succeeded
int tries; // Number of tries to transmit
const int maximumTries = 10; //Maximum number of times to try

void setup() {
  Serial.begin(9600);
  radio.begin();
  radio.openWritingPipe(pipe);
}// End setup


void loop() {
```

```
if (Serial.available() > 0) {
  dataArray[0] = Serial.parseInt();
  dataArray[1] = Serial.parseInt();
  dataArray[2] = Serial.parseInt();
  transmitData();
  if (success) {
   Serial.print("Success on attempt ");
  }
  else{
   Serial.print("Failure after attempt ");
  }
  Serial.println(tries);
}
} // End main loop

void transmitData(){
   success = 0;
   tries=0;
   while (tries<maximumTries && success==0){
   success = radio.write(dataArray, sizeof(dataArray));
   tries = tries + 1;
  }
}
```

Sketch 3.1

There are improvements you can make in this code, of course. One thing you can do is have it send the data several times to make sure that it is received correctly. This is a modification you can make.

```
void transmitData(){
   success = 0;
   tries=0;
   while (tries<maximumTries && success<3){
     success = success + radio.write(dataArray, sizeof(dataArray));
     tries = tries + 1;
```

    }
}

    This will transmit the entire dataArray array three times. The receiving program can then compare the three receptions to make sure they match, and if all three do not match, it can accept any two of the three that do match. I will get into this in the chapter on the sketch for the receiver. Of course, you can make the number of transmissions more than three if you like.

    If you do transmit the data repeatedly, you may want to delay slightly between successful transmissions to make sure the receiver has processed the data received. Here is a small modification to do that.

```
void transmitData(){
   success = 0;
   int previousSuccesses = success; // variable to determine whether to delay
   tries=0;
   while (tries<maximumTries && success<3){
     success = success + radio.write(dataArray, sizeof(dataArray));
     // This part is to cause delays
     if (success > previousSuccesses) {
       delay(100);
       previousSuccesses = success;
     }
     // End of delay section
     tries = tries + 1;
   }
}
```

    The variable previousSuccesses is used to store how many times transmission has succeeded. If the current number of successes is greater than this, there is a short delay (100 milliseconds) and previousSuccesses is set to the new number of successes.

There are other libraries for the nRF24L01 that you can find, although I do recommend using the one I referenced in the previous chapter. If you use other libraries, you may find that you have to make significant changes. One difference you may find is that in some libraries, the radio.write command does not return a value, and the compiler will deliver error messages and refuse to compile when it hits the part of the code that tries to return a value for this command. If you do use one of these libraries and encounter this problem, you can use the following version of transmitData.

```
void transmitData(){
   tries=0;
   while (tries<maximumTries){
    radio.write(dataArray, sizeof(dataArray));
    tries = tries + 1;
   }
}
```

We have the program sending an entire array of integers. A possible use for one element of the dataArray array could be to identify the transmitter. For example, you might have several Arduinos in different locations. If you do that, you probably should identify which transmitter is the source within the transmission. You can simply put something like dataArray[1] = 1 or dataArray[1] = 2 in the setup routine, where the number is different for each transmitter. I will demonstrate this in future chapters. I will also be discussing data repeaters to relay the transmissions over greater distances later in this book, and there you definitely need to identify the transmitter.

There are a few other matters worth mentioning. You can change the transmitting and receiving frequency of the nRF24L01 under software control using the radio.setChannel command. There are 125 frequencies, numbered 0 through 124. The format for the command would be, for example, radio.setChannel(120) to set the channel to 120. If you are experiencing interference with your devices, you can put the radio.setChannel command in your transmitter and receiver sketches setup routine. Of course, you need to set all your transmitters and receivers to the same channel. WiFi uses the lower channels, so to avoid interference, it is generally recommended that you set the frequency to a higher frequency, around 120. The frequency returns to the default frequency when you turn power on and off again.

You can change the transmission rate of the data using the radio.setDataRate command. The lower the transmission rate, the higher your transmission range. If you want maximum range and are willing to give up some speed of getting your data, you should consider this. The speeds are 250 kbs, 1 Mbps, and 2 Mbps. The formats for these speeds are, respectively,
radio.setDataRate(RF24_250KBPS),
radio.setDataRate(RF24_1MBPS),
and radio.setDataRate(RF24_2MBPS)

You can also increase transmission power using the command radio.setPALevel(RF24_PA_MAX). Note that you should only do this if you have a powerful power supply for your nRF24L01. The standard Arduino Uno supplies only 50 mA on the 3.3 V terminal and 500 mA on the 5 V. This is one reasons I strongly recommend using the BUONO, which delivers 300 mA on the 3.3 V output and 2 A on the 5 V output. The way to get the most power is probably to use the BUONO 5 V output to power the adapter shown in Figure 1.5, although I have not found details of how much power the voltage regulator on the adapter supplies.

If you use any of these, they should be put in the setup routine after the radio.begin() command in both transmitter and receiver sketches.

Chapter 4

Software for the nRF24L01 Receiver

Now that we have the data transmitter, the next step is to design the receiver. Much of the code is similar to the transmitter. You start out by setting up the same libraries and some of the same constants and variables like this.

#include <SPI.h>
#include <RF24.h>

#define CE_PIN 9
#define CSN_PIN 10
RF24 radio(CE_PIN, CSN_PIN); // Create a Radio

const uint64_t pipe = 0xE8E8F0F0E1LL;
#define numData 3
int dataArray[numData];
int x;

Here we are defining the constant numData to hold how many variables will be received in the array. Defining this here instead of simply using 3 throughout the code has several advantages. It makes it much easier to change the number of variables later because you can just change it in this one place instead of trying to find it throughout the code. It also makes the code much easier to understand, because we will later be using many arrays, some of which have several dimensions, and this makes it easier to see where we are talking about the number of variables as supposed to something else. The variable x is a counting variable for use in for loops. You will not need the variables maximumTries, success, or tries from the transmit program.

You also need a similar setup routine.

```
void setup() {
 Serial.begin(9600);
 radio.begin();
 radio.openReadingPipe(1,pipe);
 radio.startListening();
}// End setup
```

Notice that there are two differences. First, you are opening a reading pipe instead of a writing pipe, hence the statement radio.openReadingPipe(1,pipe) instead of radio.openWritingPipe(pipe). Notice that this takes two parameters. The second difference is that you need the statement radio.startListening(). This tells the Arduino to start listening constantly for incoming signals. You did not need this with the transmitter because when transmitting, it transmits when it has something to send, not constantly like receiving.

Instead of putting the code for listening in a separate subroutine, we will put it directly into the main loop routine. This, again, is because it will be constantly listening, unlike the transmission that could conceivably be called from several points in the code. The statement checking to see if a signal has come in is radio.available(), which is similar to Serial.available(). It returns true if there is something waiting in the radio pipe. You then use radio.read to download what has been received. The function radio.read is the exact mate of radio.write, and takes the same two parameters, the name of an array and the size of this array. The entire basic code within the loop is:

```
void loop() {
 if ( radio.available() ) {
   bool finished = false;
   while (!finished)
   {
     // Fetch the data payload
```

```
    finished = radio.read(dataArray, sizeof(dataArray));
  } // End of while
  useData();
 } //End of if radio.available
} // End main loop
```

The variable finished simply makes sure that the while statement keeps reading the data until it is all read. The data received is stored in the dataArray integer array variable. At the end of the while statement (when finished = true), the sketch encounters a call to the useData subroutine, which does whatever you want the program to do with the data. We could simply put the code right there, but for the sake of structure and easy modification I have put the action on the variable in a separate subroutine. Here is some code for demonstration.

```
void useData(){
  for (x=0;x<numData-1;x++) {
    Serial.print(dataArray[x]);
    Serial.print(",");
  } // End for x
  Serial.println(dataArray[numData-1]);
} // End useData
```

This subroutine simply sends out the integers received to the serial port as text separated by commas. For example, if dataArray[0] = 1 and dataArray[1] = 2 and dataArray[2] = 3, then the program would output "1,2,3" (without the quotation marks, of course) to the serial port. Note that the for statement goes until x reaches the number of variables minus 1, and it sends to the serial port each variable followed by a comma. For an array with 3 elements, this would output elements 0 and 1. After the for loop ends, the sketch sends the final variable followed by a line feed (because we used println instead of print). Note also that you must refer to each array element by a number 1 less than the number we would normally think of as the number of the data. For example, in a three element array, the first data item is array element number 0 and the last of three pieces of data would be array element number 2.

The data sent out the serial port can be displayed on the Arduino IDE serial monitor or input to a program running on your computer. I will give some sample code for that shortly.

The entire code is given in Sketch 4.1.

```
#include <SPI.h>
#include <RF24.h>

#define CE_PIN   9
#define CSN_PIN 10
RF24 radio(CE_PIN, CSN_PIN); // Create a Radio

const uint64_t pipe = 0xE8E8F0F0E1LL;
#define numData 3
int dataArray[numData];
int x;

void setup() {
  Serial.begin(9600);
  //delay(1000);
  Serial.println("Contact");
  radio.begin();
  radio.openReadingPipe(1,pipe);
  radio.startListening();
}// End setup


void loop() {
  if ( radio.available() ) {
    bool finished = false;
    while (!finished)
    {
      // Fetch the data payload
      finished = radio.read(dataArray, sizeof(dataArray));
    } // End of while
```

```
    useData();
  } //End of if radio.available
}// End main loop

void useData(){
  for (x=0;x<numData-1;x++) {
    Serial.print(dataArray[x]);
    Serial.print(",");
  } // End for x
  Serial.println(dataArray[numData-1]);
} // End useData
```

Sketch 4.1

In Chapter 3, I provided a transmitter sketch that transmitted the data three times for verification. To use this, the code gets a little more complicated. First, you need to define some additional variables.

```
#define numData 3
int dataArray[numData];
int holdDataArray[3][numData];
int finalDataArray[numData];
bool foundGood[numData];
int count=0;
int x;
```

We have the familiar dataArray integer array. The variable holdDataArray will hold the three sets of data collected for comparison. Note that this is a 2 dimensional array. The first element, dimensioned to hold 3, is for the 3 times the data is transmitted. The second dimension is for the multiple variables that will be received. Keeping this clear is one of the reasons mentioned previously for storing the number of variables in the constant titled numData. The variable array finalDataArray is for the final values of the data after they have been confirmed. The variable foundGood is used to store true or false to indicate whether the tests found at least two of the three transmissions matched. The variable count counts the three transmissions. That is, the first time a transmission comes is, count is 0. When the second transmission comes in it is 1, and then for the third transmission the Arduino receives count is 2.

Here is the new code for receiving and testing the data.

```
void loop() {
  if ( radio.available() ) {
    bool finished = false;
    while (!finished)
    {
    // Fetch the data payload
    finished = radio.read(dataArray, sizeof(dataArray));
    } // End of while
    for (x=0;x<numData;x++) {
      holdDataArray[count][x]= dataArray[x];
    }
    count = count + 1;
  } //End of if radio.available
  if (count == 3) {
    for (x=0;x<numData;x++){
      foundGood[x] = false;
      if (holdDataArray[0][x]==holdDataArray[1][x]){
        finalDataArray[x] = holdDataArray[0][x];
        foundGood[x] = true;
      } // End first if
      if (holdDataArray[1][x]==holdDataArray[2][x]){
        finalDataArray[x] = holdDataArray[1][x];
        foundGood[x] = true;
      } // End second if
      if (holdDataArray[2][x]==holdDataArray[0][x]){
        finalDataArray[x] = holdDataArray[2][x];
        foundGood[x] = true;
      } // End third if
      if (!foundGood[x]){
        finalDataArray[x] = -9999;
      }
```

```
} // End for x
useData();
count = 0;
} // End if count
} // End main loop
```

    The first part, the while loop that inputs the radio transmissions, is the same. However, at the end of the while loop, the sketch saves the data from dataArray into the two-dimensional array holdDataArray. For the first transmission received, the first dimension (count) is 0 and the for loop loads the variables into the second dimension. On the second transmission received, count is 1 and the for loop loads the variables into the second dimension. On the second transmission received, count is 2 and the for loop loads the variables into the second dimension. For count equals 0, 1, and 2, the program exits the if (radio.available()) section and then keeps looping back to see if there is another packet of variables received. However, after 3 packets have been received and count = 3, the program enters the if (count == 3) section. Here it starts comparing the data held in three places in the holdDataArray array to find two that match. The for x loop counts through all the variables stored in the second dimension of the holdDataArray variable. First, it set the test variable foundGood to false. Then, the first if statement compares the first set of the variables to the second set. That is, it compares each variable in holdDataArray where the element of the first dimension is 0 with the variables where the second element of the first dimension is 1. For each variable, if the two instances of the variable match, the program stores that matching value in the array finalDataArray in the same position. The second if statement compares the variables in the second position of holdDataArray (where the first dimension is 1) with the variables in the third position (where the first dimension is 2). If they match, it stores the matching value in that number of finalDataArray. Note that at this point, it is quite probable that the correct variable is already stored there from the first if statement, in which case it is simply rewriting the same value into that position. This has no effect, but does no harm. The third if statement compares the variables in the third position of holdDataArray (where the first dimension is 2) with the variables in the first position (where the first dimension is 0). Again, it is quite probable that the correct variable is already stored there from the first two if statements, in which case it is simply rewriting the same value into that position. The idea is that if any one of the three transmissions has an error in one or more of the variables, the two if statements that compare that variable will fail and the data will not be written into the finalDataArray in that position. However, whichever two are correct will match and that value will be stored in finalDataArray in that position. Therefore, assuming that each variable came through correctly any two of the three transmissions, the program will find the two matching values and put that value into finalDataArray.

    Note that when the program finds two values that match, it sets that position in foundGood to true. After the three if statements that tried to find matches, the program tests foundGood for that variable to see if it is true. If it has not been set to true, this means that no two of the three transmissions is the same. This is very unlikely. However, if that happens, the if (!foundGood[x]) saves a marker value of -9999 in that position of finalDataArray. There is no particular significance to this number, it is simply a number that is nearly impossible to have been the transmitted value (actual values returned by sensors will usually be between 0 and 1023) and therefore signals that the data was corrupted in transmission. Once the data has been placed into the finalDataArray, the useData subroutine is called. This is like the useData subroutine in the previous sketch, except that it uses finalDataArray instead of dataArray. The subroutine is shown here.

```
void useData(){
  for (x=0;x<numData - 1;x++) {
    Serial.print(finalDataArray[x]);
```

```
    Serial.print(",");
  } // End for x
  Serial.println(finalDataArray[numData-1]);
} // End useData
```

The entire sketch is given here as Sketch 4.2

```
#include <SPI.h>
#include <RF24.h>
#define CE_PIN  9
#define CSN_PIN 10

RF24 radio(CE_PIN, CSN_PIN); // Create a Radio

const uint64_t pipe = 0xE8E8F0F0E1LL;
#define numData 3
int dataArray[numData];
int holdDataArray[3][numData];
int finalDataArray[numData];
bool foundGood[numData];
int count=0;
int x;

void setup() {
  Serial.begin(9600);
  Serial.println("Contact");
  radio.begin();
  radio.openReadingPipe(1,pipe);
  radio.startListening();
}// End setup

void loop() {
  if ( radio.available() ) {
    bool finished = false;
    while (!finished)
    {
      // Fetch the data payload
      finished = radio.read(dataArray, sizeof(dataArray));
```

```
    } // End of while
    for (x=0;x<numData;x++) {
      holdDataArray[count][x]= dataArray[x];
    }
    count = count + 1;
  } //End of if radio.available
  if (count == 3) {
    for (x=0;x<numData;x++){
      foundGood[x] = false;
      if (holdDataArray[0][x]==holdDataArray[1][x]){
        finalDataArray[x] = holdDataArray[0][x];
        foundGood[x] = true;
      } // End first if
      if (holdDataArray[1][x]==holdDataArray[2][x]){
        finalDataArray[x] = holdDataArray[1][x];
        foundGood[x] = true;
      } // End second if
      if (holdDataArray[2][x]==holdDataArray[0][x]){
        finalDataArray[x] = holdDataArray[2][x];
        foundGood[x] = true;
      } // End third if
      if (!foundGood[x]){
        finalDataArray[x] = -9999;
      }
    } // End for x
    useData();
    count = 0;
  } // End if count
} // End main loop

void useData(){
  for (x=0;x<numData - 1;x++) {
    Serial.print(finalDataArray[x]);
    Serial.print(",");
  } // End for x
  Serial.println(finalDataArray[numData-1]);
} // End useData
```

Sketch 4.2

There is one conceivable problem. This sketch continues to loop without using the data until count = 3. Suppose for some reason one of the three transmissions is not received (sudden burst of static, a metal object passing between the transmitter and receiver, etc.). The program would continue to hang without displaying the data. In addition, if another three transmissions is received later, the first could be interpreted as the final packet of the previous transmission, causing the program to output the data from the previous transmission, and leaving the remaining packets in the buffer, and the whole process could repeat. It might be wise to have a time limit on how long the program waits for all three transmissions to be received. This is fairly simple. First, we need some additional variables.

unsigned long startTime;
unsigned long maximumWait = 5000;

The variable startTime is the time when the first transmission is received, and maximumWait is how long the program should wait for the final transmission to be received before it quits waiting. Note that this is in millisecond (thousandths of a second), so 5000 is 5 seconds. This is a fairly arbitrary amount, but you absolutely want to be sure that it is longer than the maximum time that it could take to transmit all three data packets. In my actual tests, however, even the sketch that transmits three times and has a 100 millisecond delay gets all three transmitted in less than two seconds.

The biggest change is in the main loop. Here is the entire new main loop.

```
void loop() {
  if ( radio.available() ) {
    bool finished = false;
    while (!finished)
    {
      finished = radio.read(dataArray, sizeof(dataArray));
    } // End of while
    for (x=0;x<numData;x++) {
```

```
      holdDataArray[count][x]= dataArray[x];
    }
    count = count + 1;
  } //End of if radio.available
  if (count == 0) {startTime = millis();}
  if (count == 3 || millis()-startTime>maximumWait) {
    for (x=0;x<numData;x++){
      foundGood[x] = false;

      if (holdDataArray[0][x]==holdDataArray[1][x] || count == 1){
        finalDataArray[x] = holdDataArray[0][x];
        foundGood[x] = true;
      } // End first if

      if (holdDataArray[1][x]==holdDataArray[2][x] && count == 3){
        finalDataArray[x] = holdDataArray[1][x];
        foundGood[x] = true;
      } // End second if

      if (holdDataArray[2][x]==holdDataArray[0][x] && count == 3){
        finalDataArray[x] = holdDataArray[2][x];
        foundGood[x] = true;
      } // End third if

      if (!foundGood[x]){
        finalDataArray[x] = -9999;
      }
    } // End for x
    useData();
    count = 0;
  } // End if count
} // End main loop
```

The entire if (radio.available()) section is the same. Immediately after that is the first additional code, the statement

if (count == 0) {startTime = millis();}

The function millis is a built in Arduino function that returns the number of milliseconds since the Arduino board started running the current program. As long as the first transmission has not been received, the startTime = millis() statement keeps setting the start time to the current time. Once one transmission is received, count no longer equals 0 and startTime no longer keeps being set to the current time, so the difference between startTime and millis() starts increasing.

Now we need to start looking at situations where count may not be 3. The next change is the addition of the code to handle the case where only one of the three transmissions is received.

```
if (count == 1) {
  finalDataArray[x] = holdDataArray[0][x];
  foundGood[x] = true;
}
```

This simply ignores any error testing and puts the values from the only holdDataArray array dimension into the finalDataArray array and marks the corresponding foundGood test as true.

Then we have the code for if more than one transmission is received.

```
if (count > 1){
  if (holdDataArray[0][x]==holdDataArray[1][x]){
    finalDataArray[x] = holdDataArray[0][x];
    foundGood[x] = true;
  } // End first comparison
}
```

This is activated whether there are two or three transmissions of the data received. It compares the first and second transmissions and if they match, stores the value in the finalDataArray array and marks the foundGood element as true. This is actually exactly the same as in the previous sketch, except that it is contained in the if (count>1) condition to make sure it does not run unless at least two transmissions were received.

If all three transmissions are received, the following code runs.

```
if (count == 3){
  if (holdDataArray[1][x]==holdDataArray[2][x]){
    finalDataArray[x] = holdDataArray[1][x];
    foundGood[x] = true;
  } // End second comparison
  if (holdDataArray[2][x]==holdDataArray[0][x]){
    finalDataArray[x] = holdDataArray[2][x];
    foundGood[x] = true;
  } // End third comparison
}
```

This, again, is actually the same comparison code as in the previous sketch, except that it is contained in the if (count == 3) test to make sure that it only runs if all three transmission are received.

You can see that all we are really doing is adding a timer to make sure the Arduino accepts the data even if less than three transmissions are received, and then making sure that it only runs comparison tests on the transmissions it receives. This is one other feature we can add since we have allowed for the data to be used even without receiving all the repeated data for checking. We may as well have the sketch inform you of how many transmissions were received so you can evaluate how reliable it is. We can do this by having the useData subroutine report to you how many transmissions were received. This just requires adding the following lines of code to the useData subroutine.

```
    Serial.print(count);
    Serial.print(",");
```

The entire sketch is shown here as Sketch 4.3.

```
#include <SPI.h>
#include <RF24.h>
#define CE_PIN   9
#define CSN_PIN 10

RF24 radio(CE_PIN, CSN_PIN); // Create a Radio

const uint64_t pipe = 0xE8E8F0F0E1LL;
#define numData 3
int dataArray[numData];
int holdDataArray[3][numData];
int finalDataArray[numData];
bool foundGood[numData];
int count=0;
int x;
unsigned long startTime;
unsigned long maximumWait = 5000;

void setup() {
  Serial.begin(9600);
  Serial.println("Contact");
  radio.begin();
  radio.openReadingPipe(1,pipe);
  radio.startListening();
}// End setup

void loop() {
  if ( radio.available() ) {
    bool finished = false;
    while (!finished) {
      // Fetch the data payload
      finished = radio.read(dataArray, sizeof(dataArray));
```

```
  } // End of while
  for (x=0;x<numData;x++) {
    holdDataArray[count][x]= dataArray[x];
  }
  count = count + 1;
} //End of if radio.available
if (count == 0) {startTime = millis();}
if (count == 3 || millis()-startTime>maximumWait) {
  for (x=0;x<numData;x++){
    foundGood[x] = false;

    if (count == 1) {
      finalDataArray[x] = holdDataArray[0][x];
      foundGood[x] = true;
    }

    if (count > 1){
      if (holdDataArray[0][x]==holdDataArray[1][x]){
        finalDataArray[x] = holdDataArray[0][x];
        foundGood[x] = true;
      } // End first comparison
    }

    if (count == 3){
      if (holdDataArray[1][x]==holdDataArray[2][x]){
        finalDataArray[x] = holdDataArray[1][x];
        foundGood[x] = true;
      } // End second comparison
      if (holdDataArray[2][x]==holdDataArray[0][x]){
        finalDataArray[x] = holdDataArray[2][x];
        foundGood[x] = true;
      } // End third comparison
    }

    if (!foundGood[x]){
      finalDataArray[x] = -9999;
    }
```

```
  } // End for x
  useData();
  count = 0;
 } // End if count
} // End main loop

void useData(){
 Serial.print(count);
 Serial.print(",");
 for (x=0;x<numData - 1;x++) {
  Serial.print(finalDataArray[x]);
  Serial.print(",");
 } // End for x
 Serial.println(finalDataArray[numData-1]);
} // End useData
```

Sketch 4.3

Now that we have the data being received, what do we do with it? There are many options. You can certainly send it to the serial monitor that comes with the Arduino IDE. If you do this, you certainly want to embellish your output with some text. For example, the useData could be rewritten as

```
void useData(){
 Serial.print("Number of transmissions received = ");
 Serial.println(count);
 Serial.print("Data received from transmitter ");
 Serial.println(finalDataArray[0]);
 Serial.print("Data item 1 = ");
 Serial.println(finalDataArray[1]);
 Serial.print("Data item 2 = ");
 Serial.println(finalDataArray[2]);
} // End useData
```

This example assumes you are using the transmitter code that sends the data array three times, the receiver code in Sketch 4.3 that checks the three transmissions and has the time out feature, and that the transmitted data array contains three integers, the first of which is an identifier number for which transmitter sent the message.

Another option would be to write a windows (or whatever operating system you like) program that inputs the data and does something with it. All of the useData examples except the last one have printed the numeric data to the serial port as a string of text characters separated by commas. A computer program can easily input the string of numbers and use the data any way it wants, such as displaying it on screen in a clearer format. One minor change we may need in the useData subroutine is to add a marker to the end of the string so the program knows when it has received all the data from the Arduino. Any string of characters that will never be part of the normal data will do. Here is a revised useData subroutine, where the marker is "X*X".

```
void useData(){
 Serial.print(count);
 Serial.print(",");
 for (x=0;x<numData - 1;x++) {
  Serial.print(finalDataArray[x]);
  Serial.print(",");
 } // End for x
 Serial.print(finalDataArray[numData-1]);
 Serial.print("X*X");
} // End useData
```

There are many programming languages you can use to write the program that accepts and uses the data from the Arduino serial output. For illustration purposes, here is a sample in Visual Basic 6. This sample uses the VB 6 Microsoft Comm control MSComm1, which you add to the project by clicking on the Project tab and then the Components item from the drop down list, then checking the box for the Microsoft Comm control, then clinking on the icon that looks like a telephone and adding it to your form. The code for using it is as follows.

```
Dim Buffer As String, S As String, P As Integer
Dim Report(3) as Integer

MSComm1.CommPort = WhichPort
' WhichPort is a variable defied elsewhere containing the
number of the serial port the Arduino is using

If MSComm1.PortOpen = False Then
   MSComm1.PortOpen = True
End If

MSComm1.InputLen = 0

  Do
  DoEvents
  Buffer = Buffer & MSComm1.Input
  P = InStr(Buffer, "X*X")
  If P > 0 Then
     Buffer = Left$(Buffer, P - 1)
     Text1 = Text1 & Buffer & vbCrLf
     'Get 1st value
     P = InStr(Buffer, ",")
     S = Left$(Buffer, P - 1)
     Report(1) = Val(S)
     lblNumberReceived = S
     Buffer = Mid$(Buffer, P + 1)
```

```
'Get 2nd value
P = InStr(Buffer, ",")
S = Left$(Buffer, P - 1)
Report(2) = Val(S)
lblTransmitter = S
Buffer = Mid$(Buffer, P + 1)

'Get 3rd value
P = InStr(Buffer, ",")
S = Left$(Buffer, P - 1)
Report(3) = Val(S)
lblData1 = S
Buffer = Mid$(Buffer, P + 1)
'Get 4th value
S = Buffer
Report(3) = Val(S)
lblData2 = S
Buffer = ""
   End If
Loop
```

The Dim statement defines variables used in this routine. The variable WhichPort is dimensioned outside this routine as a global variable and is set to the number of the com port the Arduino uses by the user elsewhere in the program. The statement
MSComm1.CommPort = WhichPort
tells MSComm1 to use this port. The statements that follow until the Do statement initialize MSComm1. These will be different for different programming languages. The part of the code that is important for general use is the part within the Do loop.

DoEvents is simply a command that tells the program to take a break and go execute other activities, such as other code within this overall program and other programs. Without this, the computer would be totally locked into this loop and would appear to freeze up. It is not actually a part of getting information from the Arduino. The first part of the

loop that actually does something is Buffer = Buffer & MSComm1.Input. This is the code that accepts string characters from the com port and adds it to the end of the string variable called Buffer. The code P = InStr(Buffer, "X*X") searches for the end marker "X*X" in the Buffer string. If it finds this, the integer variable P has the location of "X*X" in the string. For example, if there are nine characters before "X*X" , P would equal 10. If Buffer does not contain "X*X", P = 0. The program then encounters the
If P>0 Then statement, and the code between this statement and the End If statement is executed.

The first thing that happens within this If statement is that the statement
Buffer = Left$(Buffer, P - 1) strips off the "X*X" by setting Buffer equal to everything in Buffer up to "X*X". Thus, if Buffer had contained "3,1,55,77X*X" it now contains "3,1,55,77". Remember that what was sent by the Arduino in our example was the value of the variable count (probably 3), followed by a comma, followed by the identifying number of the transmitter, followed by a comma, followed by a sensor value, followed by a comma, followed by another sensor value, so "3,1,55,77" would be a reasonable content for Buffer.

In the next statement, we are assuming that the program form contains a text box called Text1. The statement Text1 = Text1 & Buffer & vbCrLf adds the string in Buffer to the contexts of the text box (vbCrLf means carriage return and line feed.) This is exactly like what the Arduino IDE serial monitor does. If you want to avoid scrolling, you could change this to Text1 = Buffer so that the text box contains only the latest string. Of course, if all your program did was display this raw data, you might as well just use the Arduino serial monitor. However, it might be nice to include this in addition to whatever your program does with the data just to be able to see the raw data coming in.

Now we get into where the program is actually using the data. The statement
P = InStr(Buffer, ",") finds the location of the first comma, and the statement

S = Left$(Buffer, P - 1) puts the string contents before this comma into the string variable S. In our example string above, this would be "3", so S = "3". The statement
Report(1) = Val(S) puts the numeric value of S into the first dimension of integer array Report. The program can now perform mathematical operations on this value, such as comparing it to some limits or other calculations. I will not show these operations in this sample code, since they will depend on the specific meaning of the data.

The statement lblNumberReceived = S assumes that there is a label on your form named lblNumberReceived, and it sets this label to the string contents f S, so this value is now showing on the screen. The line Buffer = Mid$(Buffer, P + 1) sets Buffer to the previous contents of Buffer that appear after the first comma. For example, if Buffer had been "3,1,55,77" it will now be "1,55,77".

The program then repeats this operation for each remaining value contained in the string. The value of each number is placed in the array Report, and the string contents are displayed on labels (lblTransmitter, lblData1, and lblData2) on the form. Note that in getting the last value, there is no need to search for a comma, because there is no comma after the final number. After all numbers have been extracted from the string, Buffer is set to empty (Buffer = ""), so that the Do loop can then repeat and wait for another input from MSComm1.

You might notice that there is no exit from the Do loop. You can certainly include one by changing the Do statement to Do Until expression or the Loop statement to Loop Until expression. As it is, the Do loop will continue running until you terminate the program.

This code is provided to demonstrate a basic method to input the string from the com port, store it in a string, and then parse the string into the individual numeric values for use by the program. I hope I have explained the meaning of each statement well enough that you can translate this basic program concept into what ever language you like to program in. The basic idea is to input the data string from the serial port, test for the end marker, strip the end marker away, then separate each number by searching for the commas and either display the string or convert it to a numeric value and use it.

You can also display the data on an LCD display connect to your Arduino. If you are going to do this, you need an LCD display with a serial input. Some LCD displays require six digital pins to connect, some of which are needed by the nRF24L01. A serial LCD requires only two, and these are connected to Analog inputs 4 and 5. The back of a serial LCD has a connection as shown in Figure 4.1.

Figure 4.1

You would use the same type of female-male (or female-female if you are using an Arduino with male pins) cables you used to connect the nRF24L01. The GND pin goes to the Arduino GND and the VCC pin goes to the Arduino 5V. The SDA pin goes to the A4 connection and the SCL pin connects to the A5 connection.

Here is the complete sketch for the previous receiver sketch with checking and time out using an LCD as output:

```
#include <SPI.h>
#include <RF24.h>
#include <Wire.h>
#include <LiquidCrystal_I2C.h>
```

```
// Set the LCD address to correct address for number of
columns and rows in display
LiquidCrystal_I2C lcd(0x27, 20, 4);
#define CE_PIN   9
#define CSN_PIN 10

RF24 radio(CE_PIN, CSN_PIN); // Create a Radio

const uint64_t pipe = 0xE8E8F0F0E1LL;
#define numData 3
int dataArray[numData];
int holdDataArray[3][numData];
int finalDataArray[numData];
bool foundGood[numData];
int count=0;
int x;
unsigned long startTime;
unsigned long maximumWait = 5000;

void setup() {
  Serial.begin(9600);
  lcd.begin();
  lcd.backlight();
  radio.begin();
  radio.openReadingPipe(1,pipe);
  radio.startListening();
}// End setup

void loop() {
if ( radio.available() ) {
   bool finished = false;
   while (!finished)
   {
    // Fetch the data payload
    finished = radio.read(dataArray, sizeof(dataArray));
   } // End of while
   for (x=0;x<numData;x++) {
```

```
      holdDataArray[count][x]= dataArray[x];
    }
   count = count + 1;
 } //End of if radio.available
 if (count == 0) {startTime = millis();}
 if (count == 3 || millis()-startTime>maximumWait) {
   for (x=0;x<numData;x++){
    foundGood[x] = false;

    if (count == 1) {
     finalDataArray[x] = holdDataArray[0][x];
     foundGood[x] = true;
    }

    if (count > 1){
     if (holdDataArray[0][x]==holdDataArray[1][x]){
      finalDataArray[x] = holdDataArray[0][x];
      foundGood[x] = true;
     } // End first comparison
    }

    if (count == 3){
     if (holdDataArray[1][x]==holdDataArray[2][x]){
      finalDataArray[x] = holdDataArray[1][x];
      foundGood[x] = true;
     } // End second comparison
     if (holdDataArray[2][x]==holdDataArray[0][x]){
      finalDataArray[x] = holdDataArray[2][x];
      foundGood[x] = true;
     } // End third comparison
    }

    if (!foundGood[x]){
     finalDataArray[x] = -9999;
    }

   } // End for x
   useData();
```

```
      count = 0;
    } // End if count
  } //  End main loop

void useData(){
  lcd.clear()
  lcd.setCursor(0,0);
  lcd.print("count = ");
  lcd.print(count);
  lcd.setCursor(0,1);
  lcd.print("Source = ");
  lcd.print(finalDataArray[0]);
  lcd.setCursor(0,2);
  lcd.print("Data 1 = ");
  lcd.print(finalDataArray[1]);
  lcd.setCursor(0,3);
  lcd.print("Data 2 = ");
  lcd.print(finalDataArray[2]);
} // End useData
```

We have added two library includes,
#include <Wire.h>
#include <LiquidCrystal_I2C.h>
Next we have a line to configure the LCD.
LiquidCrystal_I2C lcd(0x27, 20, 4);
The values in this line will depend on your LCD display. The first number is the address, which you must get from your LCD specs. However, it is usually 0x27 or 0x20, although I have seen 0x3F and other addresses are possible. The other two numbers are the number of columns and rows on your LCD display. The only values I have seen are 20,4 and 16,2.

You usually need to add lcd.begin(); to the setup routine, although this does vary with what library your LCD uses. You will need to check the basic setup procedure for your LCD. The command lcd.backlight(); turns on the backlight.

The loop routine is the same. No changes are necessary there to use the LCD. The main changes are in the useData subroutine. The lcd.clear command clears the screen. If you do not clear the screen, there may be residual characters left on the screen if the previous text was longer than the current text. That is, when it writes t the screen, it does not automatically clear to the end of the line. The command lcd.setCursor sets the cursor position, where the first number is the column and the second is the row. Both start with 0 and can go up to the number of rows and columns minus 1. For example, for a 40 column LCD, the value of the first number can be from 0 to 39. Each of the setCursor commands in this example sets the cursor to the first column of the various rows. This is followed by an lcd.print command that prints a label, such as "count =", and then another lcd.print command that prints the value. In this example, we are assuming you want to display the number of transmissions received (count), the source of the transmission (assuming that this was the first variable transmitted), and then two sensor readings. The output on the display is shown in Figure 4.2.

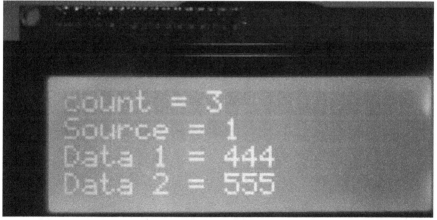

Figure 4.2

This should give you an idea of some of the ways that you can receive and display data. There is one other consideration. You might want to add sound to this too, to give you an auditory signal when a message comes in. You can do this easily in the useData subroutine. Just put a command to send a digital pin HIGH easily in the useData subroutine, like digitalWrite(1, HIGH). Then you can put a buzzer or other device on that pin. Of course, you will need to include pinMode(1, OUTPUT) in the setup. If you do not want the Arduino to signal you every time a transmission comes in, you can have some kind of test of the input conditions so that an alert sounds only when the data has certain values, like

```
If (finalDataArray[2] > 100) {
   digitalWrite(1, HIGH);
   }
else {
   digitalWrite(1, LOW);
{
```

If you have the data feed into a computer as discussed previously, you can have the computer generate a signal, such as playing a WAV file, when data comes in.

There is another option for what to do with the output from the Arduino receiver. I have written a program called Serial to Web which will post anything that comes in a serial port to a Web page. It inputs any text that comes in a designated serial port, builds it into a Web page, and uploads the Web page by FTP to a Web host. The program is called Serial to Web. It is distributed as shareware. You can download it and try it for free for 30 days. At the end of that time, if you want to continue using it, you pay $20 for a registration code that causes it to continue working after the 30-day free trial. The program can be downloaded from the Web page

https://leithauserresearch.com/s2w.html

The instructions for the program can be viewed by clicking on the Help menu at the top of the screen and then on Instructions. The instructions give considerable information on how to obtain free Web hosting. With this program, you can have the Arduino serial print anything to your computer and have Serial to Web post in online where you can view it from anywhere.

Chapter 5

Using Repeaters to Extend Your Transmission Range

As mentioned in the introduction, the nRF24L01 has a maximum range of about 1 kilometer if you use a unit with an extra antenna, assuming line of sight. You can extend this by putting repeaters between the Arduino containing the sensors and the receivers. Each repeater would receive the data, then retransmit it to the next repeater or the final receiver. The code for such a repeater is shown in Sketch 5.1

```
#include <SPI.h>
#include <nRF24L01.h>
#include <RF24.h>

#define CE_PIN   9
#define CSN_PIN 10

RF24 radio(CE_PIN, CSN_PIN);

#define numData 3
int dataArray[numData];
int holdDataArray[3][numData];
int finalDataArray[numData];
bool foundGood[numData];
int count=0;
int x;
unsigned long startTime;
int success; // variable to see if transmission succeeded
int tries; // Number of tries to transmit
const unsigned long maximumWait = 5000;
const int maximumTries = 10;
```

```
const uint64_t pipe = 0xE8E8F0F0E1LL; // Define the
transmit pipe
const uint64_t pipe2 = 0xF0F0F0F0D2LL; // Define the
receiving pipe

void setup() {
  radio.begin();
  radio.openWritingPipe(pipe);
  radio.openReadingPipe(1,pipe2);
  radio.startListening();
}// end setup


void loop() {

if ( radio.available() ) {
   bool finished = false;
   while (!finished)
   {
     // Fetch the data payload
     finished = radio.read(dataArray, sizeof(dataArray));
   } // End of while
   for (x=0;x<numData;x++) {
     holdDataArray[count][x]= dataArray[x];
   }
   count = count + 1;
} //End of if radio.available
if (count == 0) {startTime = millis();}
if (count == 3 || millis()-startTime>maximumWait) {
  for (x=0;x<numData;x++){
   foundGood[x] = false;

   if (count == 1) {
    finalDataArray[x] = holdDataArray[0][x];
    foundGood[x] = true;
   }

   if (count > 1){
```

```
      if (holdDataArray[0][x]==holdDataArray[1][x]){
        finalDataArray[x] = holdDataArray[0][x];
        foundGood[x] = true;
      } // End first comparison
    }

    if (count == 3){
      if (holdDataArray[1][x]==holdDataArray[2][x]){
        finalDataArray[x] = holdDataArray[1][x];
        foundGood[x] = true;
      } // End second comparison
      if (holdDataArray[2][x]==holdDataArray[0][x]){
        finalDataArray[x] = holdDataArray[2][x];
        foundGood[x] = true;
      } // End third comparison
    }

    if (!foundGood[x]){
      finalDataArray[x] = -9999;
    }

  } // End for x
  transmitData();
  count = 0;
} // End if count


}// End main loop

void transmitData(){
   radio.stopListening();
   success = 0;
   int previousSuccesses = success;
   tries=0;
   while (tries<maximumTries && success<3){
     success = success + radio.write(finalDataArray, sizeof(finalDataArray));
     if (success > previousSuccesses) {
```

```
   delay(50);
   previousSuccesses = success;
 }
 tries = tries + 1;
 }
 radio.startListening();
} // End transmitData
```

Sketch 5.1

This is basically combining the transmitter and receiver code. I am using the most sophisticated code, the one that transmits three times for error testing and uses timeouts in case a transmission is not received, but you can use any of the transmitter-receiver setups discussed. You can see that we have defined all the variables from each of these two sketches.

The first big thing you should notice is that two pipes are defined. One is needed for receiving and one for transmitting. In the setup routine we open one for receiving and one for transmitting. The main loop is the same as the receiver, except that instead of calling useData (which displayed data in the normal receiver) after the data has been received, it calls transmitData. This is almost exactly the same transmitData subroutine from the transmitter sketch, with a few differences. It transits the finalDataArray data instead of dataArray, because the data has gone through the confirmation process when it was received. Another very important difference you must note is the addition of the command radio.stopListening() at the beginning of the subroutine and radio.startListening at the end. This is because it cannot be both receiving and transmitting data. In order for it to do the transmission process, it must first stop listening for data (shut off the receiver). Then after it has transmitted the data, it must turn the listening back on.

There is one very serious flaw in this repeater code. It will work fine as long as you have just one repeater between the main transmitter (the one with the sensors) and the main receiver. However, consider what would happen if you had two or more. By this I mean if the main transmitter and the receiver were so far apart that you needed more than one repeater, as shown in Figure 5.1.

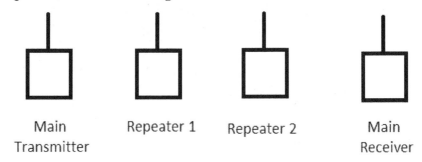

Figure 5.1

Data is transmitted from the main transmitter to repeater 1, which retransmits the data to repeater 2, which retransmits the data to the final receiver. (There could, of course, be more repeaters.) However, when repeater 2 transmits the data, repeater 1 is in range and receives it also. Repeater 1 could then interpret this as data from the main transmitter and retransmit it, which repeater 2 would then receive and retransmit, and so on. The data would keep bouncing back between repeater 1 and 2. If you had more repeaters in this string, signals would be bouncing all over the place. We need some way to tell each repeater which data to retransmit.

You remember back in Chapter 3, I mentioned that one of the values in the array you are transmitting could identify the transmitter. This is where that comes in handy. We can define a constant in each sketch like
const int myID = 1;
This would be 1 for the main transmitter, 2 for repeater 1, 3 for repeater 2, and so on. Each repeater would have a higher number. This could get a little more complicated if you have several main transmitters in range of a repeater. For example, it you had two sensor transmitters in range of repeater 1, one could be 1 and one could be 2, and then repeater 1 would be 3 and repeater 2 would be 4, but you can

work out the details. The important thing is that the ID number for the repeaters goes up as they get closer to the main receiver. Then you just add an if statement to the sketch.
if (finalDataArray[0] < myID) {transmitData();}
You would replace the transmitData() statement in the loop with this conditional statement. Of course, you could instead wrap the entire contents of the transmitData subroutine in this conditional statement if you prefer, like this:

```
void transmitData(){
  if (finalDataArray[0] < myID) {
    radio.stopListening();
    success = 0;
    int previousSuccesses = success;
    tries=0;
    while (tries<maximumTries && success<3){
      success = success + radio.write(finalDataArray, sizeof(finalDataArray));
      if (success > previousSuccesses) {
        delay(50);
        previousSuccesses = success;
      }
      tries = tries + 1;
    }
    radio.startListening();
  }
}
```

Either way, you are telling the repeater to only repeat data transmitted from the previous transmitter.

Chapter 6

Detecting Intruders

Now that we have the mechanics or the communications out of the way, let's look at some specific types of sensors we can use to supply the data. We can start with a common example, an intruder detector. There are many ways you can detect intruders. We will discuss a few.

The most common way to detect the presence of someone is with an infrared motion sensor. A typical unit is shown in Figure 6.1 and 6.2. Figure 6.2 is the better one to study because it shows the bottom of the unit where the connections are made and settings can be adjusted with the potentiometers and jumper.

Figure 6.1

Figure 6.2

As indicated in Figure 6.2, the left pin should be connected to the Arduino 5V pinhole and the right pin should be connected to the Arduino ground (GND) pin hole. The center pin is the output, which goes high (5 volts) when the sensor detects a heat source (like a human body) moving and is low normally. This can be connected to any unused Arduino digital input that is configured as input in the sketch. In the sketch in this chapter, we will use pin 4.

The delay potentiometer, on the left in Figure 6.1 and 6.2, adjusts how long the output remains high after the sensor no longer detects motion. This can usually be set from about 2 seconds to about 250 seconds. The idea is to prevent the sensor from flickering low and high whenever there is a momentary pause in the motion, but you usually want to set it low (a few seconds) so the sensor responds quickly. You can always adjust the software to disregard brief lows in the data. Setting it for short delays is accomplished by turning the potentiometer pretty close to fully counterclockwise.

The sensitivity knob has the effect of adjusting how far away a heat source can be and still be registered. Turning the knob clockwise increases sensitivity, allowing you to detect objects farther away. Maximum sensitivity is about 20 feet. Normally, you want maximum sensitivity, but you can get false positives if you set it too sensitive, so do not set it for longer range than you could possibly need. For example, do not set it for 20 feet if the sensor is only 15 feet from the opposite wall.

The jumper connects two of three pins in a row. This controls whether what is called retriggering is turned on. If retriggering is on, the output remains high as long as motion continues. If retriggering is off, the output may sometimes go low briefly even if motion continues. It is normally best for retriggering to be on. If you find that you keep getting brief lows during testing even while you are waving your hand continuously in front of the sensor, try moving the jumper to the other pair of pins.

One thing to note is that the sensor takes about 30 seconds to warming up when you first apply power. During this time, it is self calibrating and there should be no motion in front of the sensor or the calibration can be wrong. Also note that you can get some brief false positive readings during this calibration, so do not have your software do anything drastic (like call 911 for you) in the first 30 seconds after you turn on the sensor.

Now for the software. Sketch 6.1 shows the complete code for using the IR motion sensor.

```
#include <SPI.h>
#include <RF24.h>
#define CE_PIN 9
#define CSN_PIN 10
RF24 radio(CE_PIN,CSN_PIN);
```

```
const uint64_t pipe = 0xE8E8F0F0E1LL;
const int maximumTries = 10;
int dataArray[3]; // Variable to hold data
int success; // variable to see if transmission succeeded
int tries; // Number of tries to transmit
int motionSensorPin = 4;
int motionSensor;
int previousMotionSensor = LOW;
const int myID = 1;

void setup() {
  Serial.begin(9600);
  radio.begin();
  radio.openWritingPipe(pipe);
  pinMode(motionSensorPin, INPUT);
  dataArray[0] = myID;
}// End setup

void loop() {
  motionSensor=digitalRead(motionSensorPin);
  if (motionSensor != previousMotionSensor) {
    dataArray[1] = motionSensor;
    previousMotionSensor = motionSensor;
    transmitData();
  }
} // End main loop

void transmitData(){
   success = 0;
   int previousSuccesses = success;
   tries=0;
   while (tries<maximumTries && success<3){
    success = success + radio.write(dataArray, sizeof(dataArray));
     if (success > previousSuccesses) {
      delay(100);
      previousSuccesses = success;
```

```
    }
    tries = tries + 1;
  }
}
```

Sketch 6.1

This sketch is based on the transmitter sketch that transmits three times with a short delay. Instead of getting its data from you typing numbers on your computer through the serial port, it uses real data. First, we add three integer variables, motionSensorPin, motionSensor, and previousMotionSensor. The variable motionSensorPin just sets which pin inputs the reading from the motion sensor. As mentioned before, you can use any digital input pin except those already in use by the transmitter. This, incidentally, will not change in the program so you could use a constant instead of a variable. The variable motionSensor holds the value received at the motion sensor pin, and previousMotionSensor holds the previous value of MotionSensor so you can see when it has changed (as will be explained shortly). We are also using the myID constant to identify the transmitter.

In the setup routine, we have added pinMode(motionSensorPin, INPUT) to identify this pin as an input pin. This is actually not necessary, since Arduino digital pins are input by default, but it is good practice. We have also set dataArray[0] equal to myID, so the transmitter ID will be transmitted.

In the main loop, we simply set the motionSensor variable to the motion sensor input value with the statement motionSensor=digitalRead(motionSensorPin). If the value of motionSensor is not equal to what it was previously (previousMotionSensor), then we set dataArray[1] = motionSensor, set previousMotionSensor = motionSensor, and transmit the data. This causes the transmitter to only transmit when the data has changed. Of course, if you prefer, you can skip the testing and just have the transmitter send the data few seconds or so regardless of whether it has changed, with a loop statement like

```
void loop() {
    dataArray[1] = digitalRead(motionSensorPin);
    transmitData();
    delay(1000);
} // End main loop
```

However, this is not the best system. Such a system would waste power transmitting the same data over and over, and probably require you to have some code at the receiving end to alert you to when something had happened.

You might notice that we have not used the third element of the dataArray array. The value of dataArray[2] will just be 0 by default. We could change the dimensioning statement to int dataArray[2] , but I will keep the dimensions of the array at 3 in the sketches just for continuity, rather than changing the int dataArray[3] statement in each sketch.

There are a few problems with an infrared detector to detect intruders. If the intruder is wearing an insulating material, like a heavy coat, they might not show up on infrared. If they move slowly enough, their motion may not register. If the air temperature is close to body temperature, they may not be detected. Infrared motion detectors are useful devices for detecting intruders, but they are not foolproof. It might be nice to have some backup systems. Another way to detect intruders is with an ultrasonic range finder as shown in Figure 6.3.

Figure 6.3

This device sends a sound pulse out the speaker, which bounces back and is picked up by the microphone. By measuring the time it takes for the sound to come back, it can measure the distance to the nearest object in front of it. It has a maximum range of about 13 feet. Of course, at a high distance, you need a fairly large surface to bounce back the sound and register on the sensor.

By pointing this sensor at an area, you have a way of detecting the presence of an object, like an intruder. Normally, if an object (like an intruder) comes into the room, the distance measured will decrease. However, sometimes the distance might even seem to increase. Let's assume the device is set up as shown in Figure 6.4.

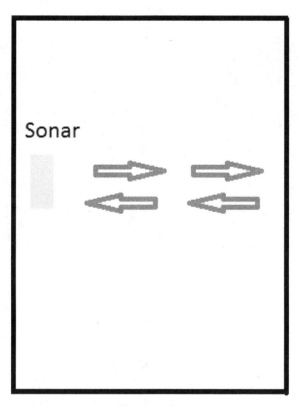

Figure 6.4

You would expect an intruder to reflect the beam as shown in Figure 6.5.

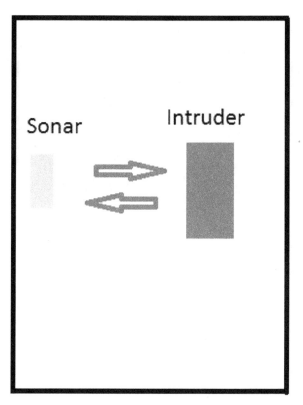

Figure 6.5

However, the intruder might deflect the beam as shown in Figure 6.6.

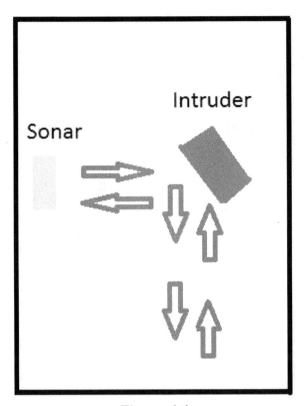

Figure 6.6

The results might be unpredictable, and the measurements imprecise. However, this does not matter. The point is that the introduction of another object into the room (or outside area) will probably change the reading, and that is all you need.

You can also point it at a door in such way that the door reflects the signal if it is opened. You can also point it at a closed door or window and allow the opening of the door or window to signal an alarm.

This device has four connections to the Arduino. There is the usual VCC connection that goes to 5V on your Arduino and GND that goes to GND for power. There is also a connection called the trigger connection (labeled Trig on the device). This goes to an output digital pin on your Arduino and is used by the Arduino to signal to send a pulse. The final connection is the echo connection that goes to a digital input pin on the Arduino to signal when the sound has been received. You can use any digital pins on the Arduino that are not already in use. We will use 7 and 8. The sketch for this is shown in Sketch 6.2.

```
#include <SPI.h>
#include <RF24.h>

#define CE_PIN 9
#define CSN_PIN 10
RF24 radio(CE_PIN,CSN_PIN);

const uint64_t pipe = 0xE8E8F0F0E1LL;
const int maximumTries = 10;
const int myID = 1;
int dataArray[3]; // Variable to hold data
int success; // variable to see if transmission succeeded
int tries; // Number of tries to transmit
int echoSensorPin = 7;
int triggerPin = 8;
long duration, distance, previousDistance;

void setup() {
  Serial.begin(9600);
  radio.begin();
  radio.openWritingPipe(pipe);
  pinMode(echoSensorPin, INPUT);
  pinMode(triggerPin, OUTPUT);
  dataArray[0] = myID;
}// End setup
```

```
void loop() {
  digitalWrite(triggerPin, LOW);
  delayMicroseconds(5);
  digitalWrite(triggerPin, HIGH);
  delayMicroseconds(10);
  digitalWrite(triggerPin, LOW);
  duration = pulseIn(echoSensorPin, HIGH);
  distance = (duration/2)/74;
  if (distance != previousDistance) {
   dataArray[1] = distance;
   previousDistance = distance;
   transmitData();
  }
  delay(50);
} //  End main loop

void transmitData(){
   success = 0;
   int previousSuccesses = success;
   tries=0;
   while (tries<maximumTries && success<3){
    success = success + radio.write(dataArray, sizeof(dataArray));
    if (success > previousSuccesses) {
     delay(200);
     previousSuccesses = success;
    }
    tries = tries + 1;
   }
}
```

Sketch 6.2

This is similar to the motion detector sketch, but a bit more complicated because the sonar sensor requires more code and does not deliver a simple HIGH or LOW output. First, we added a few variables. We set integer variables echoSensorPin = 7 and triggerPin = 8. We define variables duration, distance, and previousDistance as long (long integers). In the setup routine we need to define echoSensorPin as input and triggerPin as output.

In the main loop, the code

```
digitalWrite(triggerPin, LOW);
delayMicroseconds(5);
digitalWrite(triggerPin, HIGH);
delayMicroseconds(10);
digitalWrite(triggerPin, LOW);
duration = pulseIn(echoSensorPin, HIGH);
```

sends a short pulse out the trigger pin and waits for it to return. The function pulseIn measures the duration of a pulse. Because of the parameters given it in this code, it monitors echoSensorPin for a pulse that goes high and then low. In this case, that will be the time it took for the sound to return. This is stored in the variable duration. The statement

```
distance = (duration/2)/74;
```

converts this time interval to distance. In this case, the distance is in inches. If you prefer centimeters, replace the 74 with 29. It really does not matter, and the measurement (as mentioned before) does not even have to be precise, as long as it changes. This is handled by the code

```
if (distance != previousDistance) {
  dataArray[1] = distance;
  previousDistance = distance;
  transmitData();
}
```

As you can see, if the distance does not match the previous distance recorded, the new reading is stored in the data array, the current distance is stored in previousDistance, and the data is transmitted. Note that this will keep transmitting new data as the intruder moves around the room, and even leaves it.

A simple way to monitor doors and windows is to place a magnet on the door or window and a magnetic sensor, such as a reed switch or Hall effect sensor, on the door or window frame near the magnet. These magnetic sensors detect the presence of a magnetic field, and if the door or window is opened, the magnet is moved away from the sensor, causing a change in readings. A reed switch simply consists of two pieces of metal that are pulled together and make contact when exposed to a magnetic field. It is a simple on-off switch, which goes on when a magnetic field with the proper orientation is near. These generally output HIGH when the magnet is near. Hall effect sensors have a variable voltage output depending on the strength of the magnetic field they are exposed to (which basically means the distance from the magnet), although many have internal analog to digital circuitry that triggers a high/low output from the analog voltage. If it has an analog output, this will get either higher or lower when the magnet gets closer, depending on the orientation of the magnet and the Hall effect sensor. Some Arduino reed switches are shown in Figure 6.7, and some Hall effect sensors in Figure 6.8. You can build your own out of basic components, but these sensor boards built specifically for Arduinos are so cheap (under $1.00 from China, a few dollars from the US) that it is not worth it. The raw components are harder to find and usually end up costing just as much as a complete unit.

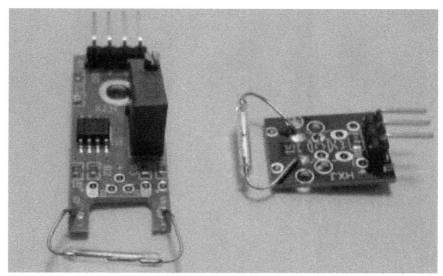

Figure 6.7

Figure 6.8

With the reed switches, the magnet should be placed parallel to the reed, as shown in Figure 6.9, for maximum range.

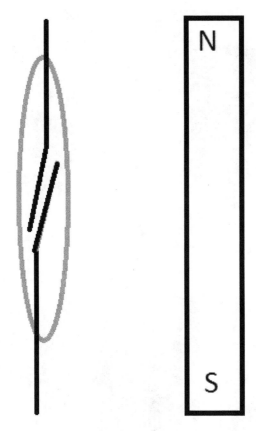

Figure 6.9

The north-south direction of the magnet does not matter with a reed switch. For Hall effect switches, this does matter, and you will need to experiment with what direction to point the magnet for maximum range.

You can set a variable or constant named digitalPin to the number of a digital input pin and initialize the input pin with pinMode(digitalPin, INPUT_PULLUP);
You can then connect one connection of the reed switch to digital input pin and the other connection of the reed switch to ground. The value of the digital input pin will then read HIGH when the reed switch is open (no magnet near it) and LOW when it is closed (magnet is near it). For extra stability, you can also connect a resistor (about 10 K) between the digital input and 5V for extra pull up. You can then monitor the reed switch with the loop code shown below.

```
void loop() {
  int sensor=digitalRead(digitalPin);
  if (previousSensor != sensor){
    dataArray[1]=sensor;
    transmitData();
    previousSensor = sensor;
  }
} // End main loop
```

This stores the reading from the digital input pin you have connected the reed switch output to in the variable sensor and compares it to the previous reading. If they are different, it stores the reading in the data array, transmits it, and then sets previousSensor to the current value of sensor. Note that this will transmit when the door or window opens and also when it is closed again. At the receiving end, you will probably want to write your code so that the alarm continues to sound until you turn the alarm off once the reading shows that the door or window has been opened. You probably do not want the alarm to shut off as soon as the door or window is closed and the Arduino sends another transmission indicating this. Having the Arduino send the transmission each time the status changes allows you to monitor the current status of the doors and widows, but having the receiver record any openings tells you if there has been a breech.

The above code will work if the sensor outputs a digital HIGH/LOW signal. If it gives you an analog signal, you need code to convert this to a yes/no answer such as that below.

```
const int trigger = 500;
bool belowTrigger = false;
bool previousBelowTrigger = false;

void loop() {
  int sensor = analogRead(A0);
  belowTrigger = false;
  if (sensor < trigger) {
    belowTrigger = true;
```

```
  }
  if (previousBelowTrigger != belowTrigger){
    dataArray[1]=belowTrigger;
    transmitData();
    previousBelowTrigger = belowTrigger;
  }
} // End main loop
```

    The constant integer trigger is a value that that Hall effect sensor will be above or below depending on whether the magnet is near the sensor. You will need to determine an appropriate value for your magnet, sensor, magnet orientation, and distance you can place the magnet from the sensor. The Boolean variable belowTrigger is used to store whether the analog sensor reading is above or below this level, and previousBelowTrigger is the previous value for this variable. The variable sensor stores the value from analog input 0 (or whichever analog input you connected the sensor to). The test variable belowTrigger is set to false, and then the if (sensor < trigger) test sets it to true if the reading is below the trigger point. Setting it to false and then setting it to true under certain circumstances is a slightly simpler way than using and if-else combination to set the variable either way. Once the value of belowTrigger is set, the code tests to see if this value is the same as the previous value. If not, the current value is stored in the second element of dataArray, the data is transmitted, and previousBelowTrigger is set to the current value of belowTrigger. As with the reed sensor code above, this transmits a signal when the door or window is opened and again when it is closed. The comments from the discussion of the reed switch about recording when the window has been opened even after it has been closed again apply here.

The final type of sensor we will describe in this chapter is electric eyes. You need only set up a narrow beam of light across the path the intruder must take and have it shine on a photocell, generally a photoresistor. A photoresistor has a resistance which decreases as the intensity of light hitting it increases. You can buy individual photoresistors and wire them into circuits yourself, or buy a ready made Arduino photoresistor sensor module. Figure 6.10 shows an individual photoresistor and a complete photoresistor sensor module.

Figure 6.10

With the module, you connect the pin on the left (as seen in the picture with the pins pointed toward you) to GND and the middle pin to 5V on the Arduino. The pin on the right goes to an analog input, such as A0. As light intensity goes down, the reading on the analog input goes up. If you use the plain photocell, you can use the circuit in Figure 6.11.

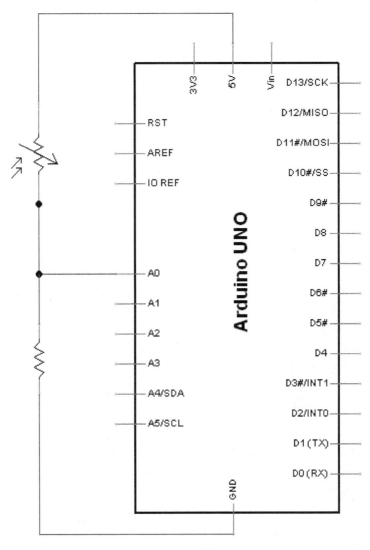

Figure 6.11

In this circuit, the analog reading will go down as light intensity decreases, meaning something has blocked the light beam. If you prefer the opposite effect, you can switch the positions of the photoresistor and the resistor. The ideal value of the resistor will vary depending on what photoresistor you use, but I find about 22K works well.

As for the sketch, whether you use the complete sensor or build your own using individual components, the loop sketch given about for the Hall effect sensor will work perfectly for the photoresistor as well, bearing in mind that you will need to set the trigger value depending on your exact components. For clarity, I will give the entire sketch here as Sketch 6.3.

```
#include <SPI.h>
#include <RF24.h>
#define CE_PIN 9
#define CSN_PIN 10
RF24 radio(CE_PIN,CSN_PIN);

const uint64_t pipe = 0xE8E8F0F0E1LL;
const int maximumTries = 10;
const int myID = 1;
int dataArray[3]; // Variable to hold data
int success; // variable to see if transmission succeeded
int tries; // Number of tries to transmit
const int trigger=500;
bool belowTrigger=false;
bool previousBelowTrigger=false;

void setup() {
  radio.begin();
  radio.openWritingPipe(pipe);
  dataArray[0] = myID;
}// End setup

void loop() {
  int sensor = analogRead(A0);
  belowTrigger = false;
  if (sensor < trigger) {
    belowTrigger = true;
  }
  if (previousBelowTrigger != belowTrigger){
    dataArray[1]=belowTrigger;
    transmitData();
    previousBelowTrigger = belowTrigger;
```

```
    }
} // End main loop

void transmitData(){
    success = 0;
    int previousSuccesses = success;
    tries=0;
    while (tries<maximumTries && success<3){
      success = success + radio.write(dataArray, sizeof(dataArray));
     if (success > previousSuccesses) {
       delay(200);
       previousSuccesses = success;
     }
     tries = tries + 1;
   }
}
```

Sketch 6.3

Chapter 7

Fire

There are several ways to detect fire. The simplest are smoke sensors, air temperature sensors, and infrared sensors tuned to the wavelengths fire will give off. I will discuss sensors for measuring air temperature in Chapter 8 and smoke (as well as other gases) in Chapter 9, so I will keep this chapter short and discuss only infrared flame detectors in this chapter.

You can purchase infrared sensors specifically designed to detect fires on various Web sites, including eBay and Amazon.com for under $3.00. Figure 7.1 shows a typical sensor. There may be some minor difference in appearance between manufacturers, but all these sensors I have found have the same basic features.

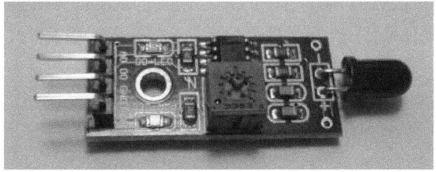

Figure 7.1

This has four pins. The center two are for power, with the center right (as seen in this picture with the pins pointed toward you) going to 5V and the center left going to GND. The right pin is digital output and the left pin is analog output. The analog output goes down as intensity of infrared light goes up. Thus, a low analog reading indicates that a flame is near the sensor. The digital pin goes from high to low when the analog reading drops to about half its top value.

(The top value is 1023 on most Arduinos, but is 4095 on some.) There is a potentiometer on the sensor that adjusts the sensitivity of the sensor. Some descriptions suggest that this adjusts the trigger point of the digital switch, but I do not find this to be correct. What it actually does is adjust how much the analog output changes with a given amount of infrared light. Turn it far clockwise and a match 5 feet away will drop the analog reading 50%. Turn it far counterclockwise and a flame a few inches away will barely move it a few percent. For most purposes, you want it to be very sensitive so that it can detect a fire on the other side of the room if you put the sensor on a wall, but if you make it too sensitive, you could get false readings. For example, sunlight can be interpreted as fire if it is too sensitive. Of course, you want to arrange the sensor so that sunlight does not fall directly onto it, or be reflected to it.

Although the flame detector does have the digital output pin, you are better off using the analog output and having your software interpret the results. This gives you more control of the interpretation. It lets you set the trigger point at just below the lowest level that the readings ever get under various light conditions short of a fire.

The sketch for this sensor is given in Sketch 7.1.

```
#include <SPI.h>
#include <RF24.h>

#define CE_PIN 9
#define CSN_PIN 10
RF24 radio(CE_PIN,CSN_PIN);

const uint64_t pipe = 0xE8E8F0F0E1LL;
const int maximumTries = 10;
const int myID = 1;
int dataArray[3]; // Variable to hold data
int success; // variable to see if transmission succeeded
int tries; // Number of tries to transmit
int outputPin = 8;
const int trigger = 800;
```

```
bool wasTriggered = false;

void setup() {
  radio.begin();
  radio.openWritingPipe(pipe);
  pinMode(outputPin, OUTPUT);
  dataArray[0] = myID;
}// End setup

void loop() {
  int sensor=analogRead(A0);
  if (sensor < trigger) {
    dataArray[1] = sensor;
    transmitData();
    digitalWrite(outputPin, HIGH);
    wasTriggered = true;
  }
  else{
   if (wasTriggered){
    dataArray[1] = 11111;
    transmitData();
    digitalWrite(outputPin, LOW);
    wasTriggered = false;
   }
  }
} // End main loop

void transmitData(){
   success = 0;
   int previousSuccesses = success;
   tries=0;
   while (tries<maximumTries && success<3){
    success = success + radio.write(dataArray, sizeof(dataArray));
    if (success > previousSuccesses) {
     delay(200);
     previousSuccesses = success;
    }
```

```
    tries = tries + 1;
  }
}
```

Sketch 7.1

I have added outputPin = 8 to provide a digital output pin in case you want to have something happen on the transmitter end, like an alarm sounding, when fire is detected. You simply connect a buzzer, LED, or other device to this pin if you want. This can also be useful for testing and calibration. The constant integer trigger allows you to set the trigger point at which the analog signal triggers a transmission. You will need to determine the proper value for this to trigger a transmission only when fire is detected, as explained previously. The Boolean variable wasTriggered allows the program to keep track of whether it has already started the reporting process.

In the main loop, the analog reading is stored in the variable sensor. If this value is less than the trigger point you set (remember that analog readings go down as infrared radiation levels go up), four things happen. First, the value of sensor is stored in the data array. The data array is then transmitted. The optional output pin is then set high. Finally, the variable wasTriggered is set to true.

The if statement has an else clause. If the sensor reading is not less than the value you selected for trigger, the code tests to see if wasTriggered is true. If not, there is no need to do anything. If wasTriggered is true, this means that fire was previously detected but now is not. In that case, the code loads the value 11111 into dataArray[1], which currently has the latest sensor reading. This is an arbitrary value that could not have come from the sensor. The data array is then transmitted. This means that the receiver will be showing 11111 and this figure will remain as the last received data. The output pin is then set to LOW, turning off whatever alarm or other indicator you had connected to this pin, if you decided to have one. Finally, the variable wasTriggered is set to false.

With this sketch, the Arduino continues to send the reading from the sensor as long as the readings are less than the trigger level. In the previous chapter, the code usually sent data only when there was a specific change in a set trigger point, like a door opening or closing. There was no value in repeatedly transmitting the fact that the door was open. In the case of the fire detector, I think it can be valuable to have it continuously transmitting a reading of the infrared radiation, so you can be seeing how the fire is progressing. If the readings do not get any lower, for example, the fire may be contained, such as in a trash can. If the readings are very steady, it can even indicate a false alarm, such as sunlight striking the sensor. If, on the other hand, the readings continue to get lower, this could indicate the fire is spreading or getting hotter. If the receiver gets the 11111 code, this tells you that the readings are no longer lower than the trigger point. Either the fire has been put out, or this was a false alarm, perhaps some electrical surge in the Arduino circuit. If you get a brief reading below the trigger point followed very quickly by the 11111 signal, it was probably a malfunction. On the other hand, if the readings suddenly stop changing while they are low and you do not get the 11111, this indicated that the Arduino has stopped transmitting, not that the fire has stopped. If the last reading indicates that he fire is still going, especially if it is a very low reading, this probably indicates that the Arduino just got burned up.

Chapter 8

Temperature Extremes

In the previous chapter, we discussed measuring infrared radiation to detect heat from a fire. In this chapter, we will talk about measuring air temperature. Extremes in air temperature can be hazardous, either too hot or too cold. Too hot can mean a fire. Even if it does not mean a fire yet, it could indicate a situation that needs attention. Heat could be rising to a point where a fire could start by spontaneous combustion. Animals could die. Equipment can overheat. On the other hand, too cold can also cause problems. Water pipes can burst, animals can freeze to death, and even plants can be harmed. Both extremes will be monitored in this chapter.

There are quite a few sensors you can use to monitor temperature. One of the best known is the DHT11, which can measure temperatures from 0 C (32 F) to 50 C (122 F). Two versions of this are shown in figure 8.1.

Figure 8.1

The one on the left is the basic component, the one on the right is the basic unit packaged with a few extra components. You will at first notice that the one on the right has four pins, while the one on the left has three. However, if you look closely, you will see that the one on the left actually has four connections going into the backing. The third pin from the left (as shown in this picture) of the four pins is not used. On the basic component, the first pin on the left goes to 5V, the second pin is the output which goes to a digital input on the Arduino, and the pin on the right goes to GND. When using this basic unit, you should connect a pull-up resistor between the output and 5V. The value should be between 4.7 K and 10K. On the three-pin package, this is built in so you do not need to provide it. On the package, which pin goes to which Arduino connection varies with the manufacturer. On the one in the picture, the output is the pin on the left, 5V is the middle pin, and the right pin goes to GND. However, I also have a package where the left pin is GND, the middle pin is output, and the right pin goes to 5V, so check your package carefully. Usually at least two are labeled. Output is often labeled S and GND as -.

Sketch 8.1 shows some code for this sensor.

```
#include <SPI.h>
#include <RF24.h>
#include <dht.h>

dht DHT;

#define inputPin 5
#define CE_PIN 9
#define CSN_PIN 10
RF24 radio(CE_PIN,CSN_PIN);

const uint64_t pipe = 0xE8E8F0F0E1LL;
const int maximumTries = 10;
const int myID = 1;
```

```
int dataArray[3]; // Variable to hold data
int success; // variable to see if transmission succeeded
int tries; // Number of tries to transmit
float tempC;
float tempF;
bool wasExtreme;
const float upperLimit = 100;
const float lowerLimit = 34;

void setup() {
  radio.begin();
  radio.openWritingPipe(pipe);
  dataArray[0] = myID;
}

void loop() {
  int error = DHT.read11(inputPin);
  if (error == 0){
    tempC = DHT.temperature;
    tempF = (tempC * 1.8) + 32.0;
    if (tempF < lowerLimit || tempF > upperLimit){
      dataArray[1] = tempF;
      dataArray[2] = DHT.humidity;
      transmitData();
      wasExtreme = true;
    }
    else{
     if (wasExtreme){
       dataArray[1] = 11111;
       transmitData();
       wasExtreme = false;
     }
    }
  }
  delay(2000);
}

void transmitData(){
```

```
    success = 0;
    int previousSuccesses = success;
    tries=0;
    while (tries<maximumTries && success<3){
      success = success + radio.write(dataArray,
sizeof(dataArray));
      if (success > previousSuccesses) {
        delay(200);
        previousSuccesses = success;
      }
      tries = tries + 1;
    }
}
```
Sketch 8.1

The first thing you need to do is download the DHT11 library. You can download the one used for the above sketch here.
https://arduino-info.wikispaces.com/DHT11-Humidity-TempSensor
or
https://arduino-info.wikispaces.com/file/detail/DHT-lib.zip
or
https://app.box.com/s/sj4d9oafq0p7b41zgxtn1w4dswhiagzi

Install the library the same way you installed the nRF24L01 libraries, as described in Chapter 2. To include the library in the sketch, we have added the line
#include <dht.h>
and the line
dht DHT;
to activate the library for the Arduino.
        The line
#define inputPin 5
sets the pin that the sensor is connected to, which can be any digital pin you want to use.

We have a few variables and constants to define. The float variables tempC and tempF will be the temperature in Centigrade (AKA Celsius) and Fahrenheit. I will have the program compute both of these just to make it easy for you to use whichever you prefer. The Boolean variable wasExtreme is used to keep track of whether the temperature was at an extreme, either high or low, on the previous pass through the loop. The float constants upperLimit and lowerLimit will be set to the temperatures you want to trigger an alert. In this example, I have set these for 100 and 34. In this sketch I will use the tempF variable for comparisons, so these temperatures are in Fahrenheit. One minor point to note: The DHT11 seems to give temperature and humidity in integer values, so using float variables may seem unnecessary. However, it does make the temperatures more accurate when you convert Celsius to Fahrenheit. Also, note that the dataArray array you are using is in integers, so when the data is received, it will be rounded to an integer anyway.

No preparations are necessary in the setup routine to use the DHT sensor. In the main loop, the sketch uses the function DHT.read11(inputPin) to read the temperature. This function returns an error code as an integer, which in this sketch is stored in the variable error. If the value is 0, there was no error and the temperature and humidity have been read correctly. Values such as -1 and -2 are returned for errors such as checksum error and timeout. Notice the if (error == 0) test. The program will only analyze the data if there was no error. If the data had an error, the loop simply repeats until good data is obtained.

The statement tempC = DHT.temperature loads the temperature into the variable tempC. Note that the readings from the DHT are in Celsius. The next line computes the Fahrenheit from this and stores it in tempF. You do not have to do this if you are comfortable working in Celsius, but I am providing this for people who are not. The next line uses an if statement to determine if the temperature is below the lower limit or above the upper limit you have set for acceptable temperature. If it is, the temperature reading is loaded into the data array in preparation for transmission. In the next line, the humidity is also loaded into the data array for transmission, just in case you might find this data useful also. The next line calls the transmission subroutine, and the next line sets wasExtreme to true, indicating that the program has started transmitting data. This is used in the else clause that follows.

If the temperature is not below the lower limit or above the upper limit, the if clause is not executed and the else clause is. This starts by checking to see if the wasExtreme variable is set to true. If it is, this indicates that until now, the temperature had been in either the high or low extreme range, and the Arduino had been transmitting data. If wasExtreme is true, the program sets dataArray[1] = 11111 and transmits this, thus signaling the "all clear." It then sets wasExtreme to false so that the "all clear" will not be transmitted again as long as the temperature stays inside the acceptable range. Thus, your receiver will now be showing an indication that the temperature had been outside the acceptable range, but is no longer.

A similar temperature sensor you can use is the DHT22. This is slightly more expensive, but is more accurate, gives the temperature and humidity to one decimal place rather than as integers. Most importantly, it has a temperature range from -40 C (-40 F) to 80 C (174 F), which can be more useful than the DHT11 for measuring temperature extremes. The DHT22 looks similar to the DHT11 except that the DHT11 is usually blue while the DHT22 is usually white, and the pin connections (5V, output, and GND) are the same. From a hardware standpoint, you can swap a DHT22 with a DHT11 with no changes. However, the drivers are a little different and thus the code is too. One possible code is shown in Sketch 8.2. You can download the library used in this sketch (Arduino-DHT22-master.zip) from
https://github.com/ringerc/Arduino-DHT22
or
https://app.box.com/s/pjt1f2vtdvgwmvvrs88hrihj8oi88vmc

```
#include <SPI.h>
#include <RF24.h>
#include <DHT22.h>

#define CE_PIN 9
#define CSN_PIN 10
RF24 radio(CE_PIN,CSN_PIN);
#define DHT22_PIN 2
DHT22 myDHT22(DHT22_PIN);

const uint64_t pipe = 0xE8E8F0F0E1LL;
const int maximumTries = 10;
const int myID = 1;
int dataArray[3]; // Variable to hold data
int success; // variable to see if transmission succeeded
int tries; // Number of tries to transmit
float tempC;
float tempF;
const float upperLimit = 100;
const float lowerLimit = 34;
void setup() {
 Serial.begin(9600);
 radio.begin();
 radio.openWritingPipe(pipe);
 dataArray[0] = myID;
}

void loop() {
 DHT22_ERROR_t error;
 error = myDHT22.readData();
 if (error == 0){
  tempC = myDHT22.getTemperatureC();
  tempF = (tempC * 1.8) + 32.0;
  if (tempF < lowerLimit || tempF > upperLimit){
   dataArray[1] = tempF;
   dataArray[2] = myDHT22.getHumidity();
   transmitData();
  }
```

```
  }
  delay(2000);
}

void transmitData(){
   success = 0;
   int previousSuccesses = success;
   tries=0;
   while (tries<maximumTries && success<3){
    success = success + radio.write(dataArray, sizeof(dataArray));
    if (success > previousSuccesses) {
      delay(200);
      previousSuccesses = success;
    }
    tries = tries + 1;
   }
}
```

Sketch 8.2

The functioning of this code is exactly the same as Sketch 8.1 for the DHT11. The only changes are the exact wording of the statements to accommodate the different library. For example DHT.temperature is replaced with myDHT22.getTemperatureC()
and int error = DHT.read11(inputPin) is replaced by the two lines DHT22_ERROR_t error and error = myDHT22.readData(). Of course, the include statement #include <dht.h>has been changed to #include <DHT22.h> and dht DHT has been changed to
DHT22 myDHT22(DHT22_PIN). You might notice that the delay has been increased from 1000 milliseconds to 2000 milliseconds. This sensor takes a little longer to read, probably because of the greater accuracy.

Chapter 9

Toxic and Flammable Gases

There are a variety of gas sensors that can be attached to the Arduino to detect quite a few different gases. Some of these sensors detect one gas. Others detect several, and it is not always easy to determine which gas is being detected purely from the readings. These sensors usually have a designation that starts with MQ, followed by a number which indicates which gases they detect. Table 9.1 on the next page lists the most commonly available sensors.

For simplicity, I will deal only with the complete packages that you can buy for Arduinos. If you use the raw component, you need to provide pull up resistors and voltage regulators and do a lot of extra wiring. These sensors have an analog output, and the output voltage will go up or down depending on the amount of gas present. As far as I can determine, the analog output always goes up as the gas concentrations increase, although I have not tested every sensor.

Figure 9.1 shows some of these sensors. They all look pretty much the same.

Figure 9.1

These devices have four pins. These are VCC (which connects to 5 V on the Arduino) and GND power pins, an analog output pin (AOUT), and a digital pin (DOUT). The digital pin usually switches from high to low when the analog output crosses a threshold that you can usually set with a built-in potentiometer, but in some cases the digital pin allows you to adjust the heater that all of them have. The arrangement of the pins varies with the manufacturer, so I will not attempt to describe them here.

Name	Gases	Notes
MQ2	methane, butane, LPG, smoke	LPG is liquefied natural gas, a combination of propane and butane
MQ3	alcohol, ethanol, smoke	Alcohol can be used as crude breath tester

MQ4	methane, CNG (compressed natural gas)	Natural gas is mostly methane with impurities
MQ5	natural gas, LPG	
MQ6	LPG, butane	
MQ7	CO (carbon monoxide)	Need to cycle power to heater between 5V and 1.4V
MQ8	Hydrogen gas	
MQ216	Natural gas, coal gas	Coal gas is hydrogen, methane, and carbon monoxide
MQ303A	Alcohol, ethanol, smoke	
MQ306A	LPG, propane	
MQ131	Ozone	Requires 6V power supply
MQ135	ammonia, benzene, alcohol, smoke	
MQ136	H2S (hydrogen sulfide)	
MQ137	ammonia	
MQ138	Benzene, toluene, alcohol, acetone, propane, formaldehyde gas, hydrogen	
MG811	CO2 (carbon dioxide)	Requires 6V power supply, best used with op-amp

Table 9.1

Since these have an analog output, the same basic sketches we have used before (such as 7.1) will work with these. The primary difference is that with the sensor for Sketch 7.1, the analog readings went down when a fire was detected, whereas for these sensors the analog reading go up when gas is detected. The basic sketch for most of these sensors is given in Sketch 9.1

```
#include <SPI.h>
#include <RF24.h>

#define CE_PIN 9
#define CSN_PIN 10
RF24 radio(CE_PIN,CSN_PIN);

const uint64_t pipe = 0xE8E8F0F0E1LL;
const int maximumTries = 10;
const int myID = 1;
int dataArray[3]; // Variable to hold data
int success; // variable to see if transmission succeeded
int tries; // Number of tries to transmit
int outputPin = 8;

const int trigger = 100;
bool wasTriggered = false;

void setup() {
  radio.begin();
  radio.openWritingPipe(pipe);
  pinMode(outputPin, OUTPUT);
  dataArray[0] = myID;
}// End setup

void loop() {
  int sensor = analogRead(A0);
  if (sensor > trigger) {
    dataArray[1] = sensor;
    transmitData();
```

```
      digitalWrite(outputPin, HIGH);
      wasTriggered = true;
    }
    else{
     if (wasTriggered){
      dataArray[1] = 11111;
      transmitData();
      digitalWrite(outputPin, LOW);
      wasTriggered = false;
     }
    }
} // End main loop

void transmitData(){
   success = 0;
   int previousSuccesses = success;
   tries=0;
   while (tries<maximumTries && success<3){
     success = success + radio.write(dataArray, sizeof(dataArray));
    if (success > previousSuccesses) {
      delay(200);
      previousSuccesses = success;
    }
    tries = tries + 1;
   }
}
```

Sketch 9.1

As I have done in the past, the sketch has statements to include libraries for the radio transmitter and the radio is initiated. I assigned an ID (in this case, 1) to the transmitter and this value is stored in dataArray[0]. I have included an output pin that goes high when the sensor reading gets above the trigger point, in case you want to attach a signal (like a buzzer) to the transmitter in addition to having it transmit the signal.. Once the analog signal gets above the threshold, the Arduino repeatedly transmits the value. Incidentally, you can put a delay in the loop, possibly at the end of the if (sensor > trigger) clause, to slow down how often it sends the data.

Some sensors need extra attention because they need to alternate the heater temperature between a high (hotter) and low state. For example, the MQ-7 sensor alternates between 5V and 1.4V, and only transmits data for CO (carbon monoxide) levels while it is in the low state. As I understand it, the reason for periodically raising the temperature is to burn off flammable gases that might accumulate on the sensor. The MQ9 needs to cycle power to heater between 5V and 1.5V. It detects CO when the heater is at 5V and flammable gases when the temperature is 5V. The MQ309A does the same thing, but alternates between .2V and .9V. Sketch 9.2 is for the MQ7, and handles the cycling of the temperature and makes sure that it only takes readings at the lower temperature.

```
#include "Arduino.h"
#include <SPI.h>
#include <RF24.h>

#define CE_PIN 9
#define CSN_PIN 10
#define LED_PIN 4
RF24 radio(CE_PIN,CSN_PIN);

const uint64_t pipe = 0xE8E8F0F0E1LL;
const int maximumTries = 10;
const int myID = 1;
const int outputPin = 4;
const int upperLimit = 200;
int dataArray[3]; // Variable to hold data
int success; // variable to see if transmission succeeded
int tries; // Number of tries to transmit
bool wasTriggered = false;

class CS_MQ7{
  public:
```

```cpp
    CS_MQ7(int CoTogPin, int CoIndicatorPin);
    void CoPwrCycler();
    boolean CurrentState();
    unsigned long time;
    unsigned long currTime;
    unsigned long prevTime;
    unsigned long currCoPwrTimer;
    boolean CoPwrState;

  private:
    int _CoIndicatorPin;
    int _CoTogPin;
};

CS_MQ7::CS_MQ7(int CoTogPin, int CoIndicatorPin){

  pinMode(CoIndicatorPin, OUTPUT);
  pinMode(CoTogPin, OUTPUT);
  _CoIndicatorPin = CoIndicatorPin;
  _CoTogPin = CoTogPin;
  time = 0;
  currTime = 0;
  prevTime = 0;
  currCoPwrTimer = 0;
  CoPwrState = LOW;
  currCoPwrTimer = 500;
}

//   Subroutine to cycle power to heater between 5V and 1.4V
void CS_MQ7::CoPwrCycler(){
  currTime = millis();
  if (currTime - prevTime > currCoPwrTimer){
    prevTime = currTime;

    if(CoPwrState == LOW){
      CoPwrState = HIGH;
      currCoPwrTimer = 60000;  //60 seconds at 5v
```

```
    }
    else{
      CoPwrState = LOW;
      currCoPwrTimer = 90000;  //90 seconds at 1.4v
    }
    digitalWrite(_CoIndicatorPin, CoPwrState);
    digitalWrite(_CoTogPin, CoPwrState);
  }
}

// Report whether heater is high or low
boolean CS_MQ7::CurrentState(){
  if(CoPwrState == LOW){
    return false;
  }
  else{
    return true;
  }
}

CS_MQ7 MQ7(2, 3); // pin 2 connected to heater control from sensor board
              // 3 = digital Pin connected to LED Power Indicator

int CoSensorOutput = 0; // Pin connected to "out" from sensor board
int CoData = 0;        //analog sensor data

void setup() {
  radio.begin();
  radio.openWritingPipe(pipe);
  dataArray[0] = myID;
}

void loop() {
  MQ7.CoPwrCycler();
```

```
  if(MQ7.CurrentState() == LOW){  //  Heater at 1.4v, read sensor data!
    CoData = analogRead(CoSensorOutput);
   if (CoData > upperLimit) {
    dataArray[1] = CoData;
    transmitData();
    digitalWrite(outputPin,HIGH);
    wasTriggered = true;
   }
   else{
    if (wasTriggered){
    dataArray[1] = 11111;
    transmitData();
    digitalWrite(outputPin,LOW);
    wasTriggered = false;
   }
  }
 }
}

void transmitData(){
   success = 0;
   int previousSuccesses = success;
   tries=0;
   while (tries<maximumTries && success<3){
     success = success + radio.write(dataArray, sizeof(dataArray));
     if (success > previousSuccesses) {
       delay(200);
       previousSuccesses = success;
     }
     tries = tries + 1;
   }
}
```

Sketch 9.2

This code switches the power between high and low, and only takes readings when the power is low. As mentioned before, the purpose of the high voltage to the heater seems to be to burn off any combustible residue that could accumulate and interfere with the readings when the power is low.

There are several things to consider about these sensors. First, they are not to be considered precision sensors. They are inexpensive devices to give relative ideas of gas concentrations. They are more for detection than precise measurement, and thus are mostly good for alarms.

That said, you might want to have at least some idea of what the readings indicate, so some measurements might be helpful. Ideally, you would expose the sensor to known concentrations of gases and record the readings for each known concentration. However, that would require that you have a device to measure the precise concentration of the gas during the calibration, which would be expensive. A simpler solution is to get a reading for approximately 0% and 100% concentration of the gas and extrapolate from there. Getting the 0% reading should be pretty easy. Just put the device in an environment that presumably is clean. Often, this means just taking it outside. If you suspect that your outside air may be contaminated (like maybe you live in an area where there is a lot of fracking going on), you can spray the sensor with gas from an oxygen or nitrogen tank. If this is not practical, you could put the sensor in a small enclosed space with a high quality air purifier, preferably one with a carbon filter. The point is to get as clean an environment as possible and note the reading in that environment as your zero point. For the high point, you can spray the sensor directly with gas from a source like a methane or butane or propane tank, a gas stove, or even a gas cigarette lighter (depending on the type of gas your sensor detects). You can even put the sensor inside a plastic bag, squeeze out as much air as possible, and then inject the gas into the bag to try to get a high concentration. You will then have a low and high point to estimate the concentration for readings in between.

Before you calibrate or even use a new sensor, you should run a new sensor for about 12 hours. This burn-in period will stabilize the sensor readings. I believe that part of this is to allow the heater to burn off any residue left on the sensor during manufacture, although there may be other related effects. When you first power up the sensor each time you turn on the device, let it warm up for at least about 30 seconds before you start taking the readings seriously. You can get all sorts of wild readings in the first 30 seconds before the sensor stabilizes. I should also mention that the sensor readings can be affected by other factors, such as air temperature and humidity. You might want to combine the gas sensor with a temperature and humidity sensor and see if you find any correlation between these other factors and the gas readings you get. For example, if you find the gas readings always seem to go up as the humidity goes up, you might want to make allowances for this, such as adding some code to subtract a few points from the gas reading for a certain amount of humidity rise.

On this subject, this is a good time to mention that you can certainly combine several sensors in one device. Remember that the transmit code is sending an entire array of values. You can put the readings of one sensor in each of these array elements when you transmit the array. Sometimes the information can even be combined in ways to provide more information than the individual pieces of information. For example, the MQ2 is sensitive to methane, butane, and smoke. The MQ4 is sensitive to methane. The MQ6 is sensitive to butane. If you get a reading from the MQ2 but not the MQ3 or the MQ6, you can assume that the MQ2 detected smoke, thus you have a smoke detector. If you get a reading from the MQ2 and the MQ4, you can assume that the sensors are detecting methane. Adding other sensors, like the MQ7 and MQ8, gives you additional information. Sketch 9.3 demonstrates how to have multiple sensors in one package.

```
#include <SPI.h>
#include <RF24.h>

#define CE_PIN 9
#define CSN_PIN 10
RF24 radio(CE_PIN,CSN_PIN);

const uint64_t pipe = 0xE8E8F0F0E1LL;
const int maximumTries = 10;
const int myID = 1;
int dataArray[3]; // Variable to hold data
int success; // variable to see if transmission succeeded
int tries; // Number of tries to transmit
int outputPin = 8;

const int trigger1 = 200;
const int trigger2 = 350;
bool wasTriggered = false;

void setup() {
  radio.begin();
  radio.openWritingPipe(pipe);
  pinMode(outputPin, OUTPUT);
  dataArray[0] = myID;
}// End setup

void loop() {
  int sensor1 = analogRead(A0);
  int sensor2 = analogRead(A1);
  if (sensor1 > trigger1 || sensor2 > trigger2) {
    dataArray[1] = sensor1;
    dataArray[2] = sensor2;
    transmitData();
    digitalWrite(outputPin, HIGH);
    wasTriggered = true;
  }
  else{
   if (wasTriggered){
```

```
    dataArray[1] = 11111;
    dataArray[2] = 11111;
    transmitData();
    digitalWrite(outputPin, LOW);
    wasTriggered = false;
   }
  }
} // End main loop

void transmitData(){
   success = 0;
   int previousSuccesses = success;
   tries=0;
   while (tries<maximumTries && success<3){
    success = success + radio.write(dataArray, sizeof(dataArray));
    if (success > previousSuccesses) {
      delay(200);
      previousSuccesses = success;
    }
    tries = tries + 1;
   }
}
```

Sketch 9.3

The only changes I had to make were to add a second trigger level constant, a second sensor variable reading another analog input, change the if (sensor > trigger) to if (sensor1 > trigger1 || sensor2 > trigger2), have the second sensor reading loaded into dataArray[2] before transmission, and have 11111 loaded into dataArray[2] when both sensor readings drop below their respective trigger points. Note that the data is transmitted if either sensor goes above their own trigger point and the "all clear" value of 11111 is transmitted only when both sensors go back below their trigger points. You can, of course, apply this to any number of sensors, up to the number of inputs your Arduino has. Just add more trigger levels and sensor variables. (You will have to increase the dimensioning of the dataArray array.) This does not have to be limited to more gas sensors, too. You can add any type of sensor, such as a temperature sensor. You do not, incidentally, absolutely have to have all the sensors you add trigger a transmission. You can add various types of data, such as temperature and humidity, to the array when data is transmitted but not include those in the if statement that determines when data is transmitted.

Chapter 10

Flooding or Too Little Water

This chapter is about water detection. You might think that you are unlikely to experience a flood, but remember that this can include anything from a flood outside to heavy rain leaking into your basement or a pipe bursting. Even a small leak from a water pipe can cause damage. I should also mention that the same sensor that can detect flooding (too much water) can also alert you if water levels drop too low, such as water in your pool or an animal's water dish. For example, on a farm, you could use these sensors to alert you if water was not being supplied in an animal's water trough.

Detecting water levels is very simple. All you really need is to dangle some wires in the water separated by a short distance and measure the amount of current flowing between the wires. The more current flowing, the deeper the wires are into the water. You can easily make your own water sensor, but it is hardly worth your while. You can buy them for under a dollar on EBay. Figure 10.1 shows a typical water sensor.

Figure 10.1

As you can see, it has three connector pins, for V+, GND, and analog output (marked S). It also has rows of wires, with alternate wires being connected. When the sensor gets wet, current is conducted between each alternate wire, causing the voltage at the output pin to go up. If you want to detect the presence of any water at all, you can lay it almost flat. (If you do that, you should insulate the contacts, perhaps by putting tape over them.) If you want to detect water levels, you can stand it upright.

The analog reading when the sensor is dry is 0. Unfortunately, the readings are very nonlinear. In my tests, the moment I dipped the end in water about 2 mm deep, the reading shot up to about 400. Totally immersing it only raised the reading to about 540. After removing the sensor from the water, enough water clung to the sensor to keep the reading at about 250. A single drop lying across any two wires will keep a reading in the hundreds. Since this is based in how much current is conducted, the readings will also be influenced by the conductivity of the water. For example, if you add some salt to the water, the reading will go up at the same level of water. Thus, this can also be used as a very crude salinity meter if you have it at a steady water level.

Since this is an analog output, many of the sketches from previous chapters will work fine for this sensor. For example, Sketch 9.1 will work with this as a water detector. Of course, you will want to change the trigger point. The trigger point is not hard to adjust, since this sensor reads 0 when dry. I would set the trigger point at about 50 if all you wanted is to detect any water at all. Note that if you want to detect low water level instead of the presence of water, you need to revise the if (sensor > trigger) statement to read if (sensor < trigger) so that it detects the water level going down instead of up.

Chapter 11

Power Failure or Low Power

In this chapter I will discuss detecting power failure and also low power. At first glance a power failure might not seem like a hazard, but consider the possibility of having a power failure in a detached structure, like a barn on a farm. Serious damage could occur if you are not alerted. Also, consider that fact that if you have the Arduino/sensor unit powered by a battery, you might want to be alerted if the battery was getting low before the unit failed altogether.

First, let's consider the simpler case where you want to know if you AC power has failed. Figure 11.1 shows a simple circuit to alert you to AC power failure.

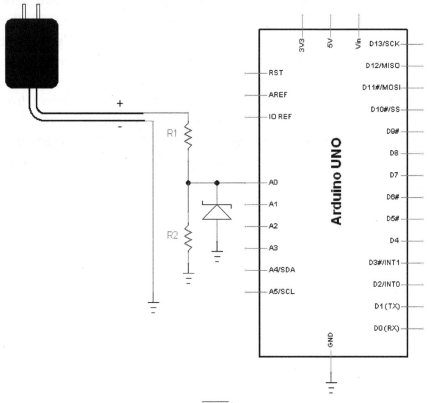

Figure 11.1

Here we have an AC to DC adapter plugged into the wall, with the negative going to the Arduino ground and the positive going to a voltage divider that is linked to the Arduino's A0 pin. The power adapter can be any power adapter that converts line AC to any DC. You can use an old one from a discarded appliance, or buy a universal adapter from most electronics stores. The purpose of the voltage divider is to reduce that DC voltage to no more than 5 V for the analog pin. The Zener diode is to guarantee that the voltage into A0 is never above 5 V, in case there is a voltage surge from the adapter or the resistors are not the values they are reported to be. Unfortunately, it is hard to find a 5 V Zener diode. You can, however, easily find a 4.7 V Zener diode, so we will aim for a maximum of 4.7 V from the voltage divider.

The formula therefore for selecting the values of R1 and R2 is
Rmin = R1 * (V/4.7 - 1)
where V is the output voltage of the adapter and Rmin is the MINIMUM value of R2. For example, if the output voltage of the adapter is 9 V and you choose 10K for R1, then the minimum value for R2 would be 9.1K, so another 10K resistor would work fine for R2. In practice, you should actually try to set the voltage at A0 a bit below 4.7 V for safety, since actual resistors vary a bit from their labeled values, so your actual R2 could be a bit lower than advertised and R1 might be a bit higher than advertised, resulting in a voltage at A0 higher than 4.7 V, with bad results. In practice, it would be better to use a higher voltage for R1 then the formula indicates. For example, if the formula gives a value of 10K, you can use 12K. Note that the actual values of the resistors is not very important, it is the ratio that matters. However, you do not want them to be too low, or they will draw a lot of current, and you do not want them to be too high, or the input resistance of the Arduino itself will come into play. I recommend using 10K for R1 and calculating R2 from there.

Now for the software. In the event of a complete power failure, the voltage at A0 would drop to 0, so all you have to do is have the sketch send an alert if the voltage drops below some arbitrary point between the normal voltage at A0 and 0. Sketch 11.1 does this.

```
#include <SPI.h>
#include <RF24.h>

#define CE_PIN 9
#define CSN_PIN 10
RF24 radio(CE_PIN,CSN_PIN);

const uint64_t pipe = 0xE8E8F0F0E1LL;
const int maximumTries = 10;
const int myID = 1;
int dataArray[3]; // Variable to hold data
int success; // variable to see if transmission succeeded
int tries; // Number of tries to transmit

const int trigger = 200;
bool wasTriggered = false;

void setup() {
  radio.begin();
  radio.openWritingPipe(pipe);
  dataArray[0] = myID;
}// End setup

void loop() {
  int sensor = analogRead(A0);
  if (sensor < trigger) {
    dataArray[1] = sensor;
    transmitData();
    wasTriggered = true;
  }
  else{
   if (wasTriggered){
```

```
      dataArray[1] = 11111;
      transmitData();
      wasTriggered = false;
    }
  }
} // End main loop

void transmitData(){
  success = 0;
  int previousSuccesses = success;
  tries=0;
  while (tries<maximumTries && success<3){
    success = success + radio.write(dataArray, sizeof(dataArray));
    if (success > previousSuccesses) {
      delay(200);
      previousSuccesses = success;
    }
    tries = tries + 1;
  }
}
```

Sketch 11.1

As I am sure you have noticed, this is basically Sketch 9.1 with if (sensor > trigger) replaced by if (sensor < trigger). When the power goes off, the reading at A0, and thus the value of the sensor variable, will drop to 0. The value of A0 when the power is on will be about 900, so any value between 0 and 900 (preferably at the lower end) will do for trigger.

Suppose you want the Arduino to be plugged into the wall to provide power to it during normal operation. This leads to the interesting question of how the Arduino can send a "no power" alert if it has no power. One answer is to use a portable USB charger such as the one shown in Figure 11.2.

Figure 11.2

These charge up by plugging them into a USB port. This can be a computer, but it can also be an AC to USB adapter, such as the one shown in Figure 11.3.

Figure 11.3

The normal use for a portable USB charger is to charge it up from a stationary source, like the AC to USB adapter, then unplug it and take it with you to use to give your phone or other device a quick charge. However, you can also plug the portable charger into the AC to USB adapter, then plug your Arduino into the portable USB charger, and you have a simple and inexpensive uninterruptible power supply (UPS) for your Arduino. Your Arduino can then monitor your AC power with the circuit shown in Figure 11.1 and signal you if there is a power outage. Of course, you could plug your Arduino into a standard UPS like you probably use for your computer, but those are much more expensive.

Now consider the situation where your Arduino is powered by a battery, and you want to monitor the battery and have the Arduino signal you if the battery is getting low. You can use the circuit shown in Figure 11.4.

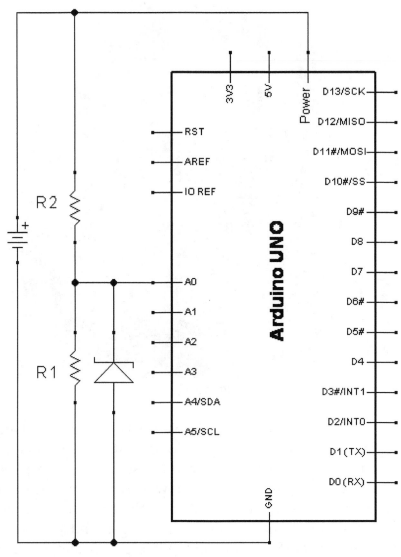

Figure 11.4

Here we have the Arduino being powered by a battery. The negative terminal of the battery goes to the Arduino GND and the positive terminal goes to the Arduino VIN connection. Note that this is NOT the 5V connection. Connecting the battery, assuming it is more than 5V, to the 5V connection would damage or destroy the Arduino. The VIN connection has a voltage regulator that will automatically drop the voltage to 5V. The rest of the circuit is basically the same as Figure 11.1 The voltage divider reduces the voltage to something under 4.7 V, and the optional 4.7 V Zener diode protects against any chance of the voltage exceeding this.

The sketch for this is basically the same as Sketch 11.1, with one important exception. In Sketch 11.1, the precise value of the trigger point was not important. Anything above 0 would work. In this case, we want to be warned when the battery voltage drops below a certain level that is much closer to the full voltage of the battery. As a battery weakens, the current it can deliver starts to drop, which results in a voltage drop in the final stages of battery failure. For this warning, we want to be alerted as soon as the voltage drops significantly. This means that we want to set trigger slightly below the reading that we get when the battery is fully charged. This is especially important if the battery voltage is only slightly above the minimum voltage to power your Arduino, because if the voltage drops below the minimum level to power the Arduino before it reaches the trigger point, it will not be able to send the alert that the power is low. Most Arduino Unos and Megas have a minimum voltage of 7 V to power them. The Arduino Buonos have a minimum of 6 V, which is one of their advantages. Let's say you are powering this from a 9 V battery. You will probably want to have the device alert you if the power goes below 8 V.

The voltage at A0 = Vb * R1/(R1 + R2), where Vb is the battery voltage. Let's say that R1 is 10K and R2 is 12K. At 9 V, the voltage at A0 would be about 4.1 V for Vb = 9 V and about 3.6 V for Vb = 8 V. Assuming your analog input reads 1023 at 5 V (some read 4095), the reading if your battery drops to 8 V would be 1023 * 3.6/5 = 737. This should be your trigger point. In general, your trigger point for this sketch should be

Vb * R1/(R1 + R2) * 1023/5

where Vb is the battery voltage at which you want to trigger the alert. Of course, if you have an Arduino that reads 4095 at 5 V, use 4095 instead of 1023 in the above equation.

Chapter 12

Other Sensors

I have discussed a few possible sensors that you could connect to your Arduino to alert you to various conditions. These are certainly no the only ones you can use. In this chapter, I will simply list a few other sensors that are designed to work with Arduinos that are easily available on EBay or Amazon.com, as well as other sources.

UV (ultra violet) - alert you when UV levels are high
Soil Hygrometer - alert you when your plants are drying out
pH - monitor the pH of your pool, Aquaponic garden, fish tank, or other water supply
sound - used as a possible additional intruder detector
vibration or tilt - used to detect movement of objects, possible earthquake detector or alert if objects are taken
other types of temperature sensors, some of which are submersible
heartbeat/pulse
barometric pressure
water conductivity - measures water salinity

In addition to being able to buy sensors specifically designed for Arduinos, you can build your own Arduino sensors using any type of sensor, even if it was not designed for Arduinos. Almost any sensor works by changing either output voltage or resistance in response to whatever it is measuring. To learn to connect these to your Arduino, I recommend another book I have written entitled "Building Your Own Arduino Shields: Interfacing with the Arduino Using Basic Components." This book is available on Amazon.com for $3.99 on Kindle or $7.99 paperback.

Appendix

Downloading files and Contacting the Author

In order to save you typing, I have made the numbered sketches in this book available for download in a ZIP file. If you download and unzip the file (Windows has an unzip function), you will get a directory named RemoteSensors. Inside this directory is a ReadMe.txt file and another folder named examples. If you move the entire RemoteSensors folder into your Documents/Arduino folder (created when you installed the Arduino IDE), the files will show up when you click on File/Sketchbook in the Arduino IDE. The files have names like S4P1, which stands for Sketch 4.1. Thus, you can load a named file like Sketch 4.1 without needing to type in the entire sketch.

I have made the ZIP file available at the following locations:
https://github.com/DavidLeithauser/remote-sensors-Arduino/tree/master
This is a public download site for programming related files. It is well established, and will probably remain in existence for a long time. To download the zip, click on the "Clone or download" button and then on "download zip."
https://LeithauserResearch.com/RemoteSensors.zip
This is my software company Web site, run at my own expense. It is currently paid up and set to remain open until at least February of 2020. I plan to maintain it beyond that, but I make no guarantees.

If you need to contact me, use email address Leithauser@aol.com with subject line Remote Sensor book.

CPSIA information can be obtained
at www.ICGtesting.com
Printed in the USA
BVHW040804250420
578457BV00014B/2912